Springer Tracts in Advanced Robotics

Volume 139

Frank Chongwoo Park, Mechanical Engineering Department, Seoul National University, Seoul, Korea (Republic of)

S. E. Salcudean, The University of British Columbia, Vancouver, BC, Canada

Roland Siegwart, LEE J205, ETH Zürich, Institute of Robotics & Autonomous Systems Lab, Zürich, Switzerland

Gaurav S. Sukhatme, Department of Computer Science, University of Southern California, Los Angeles, CA, USA

The Springer Tracts in Advanced Robotics (STAR) publish new developments and advances in the fields of robotics research, rapidly and informally but with a high quality. The intent is to cover all the technical contents, applications, and multi-disciplinary aspects of robotics, embedded in the fields of Mechanical Engineering, Computer Science, Electrical Engineering, Mechatronics, Control, and Life Sciences, as well as the methodologies behind them. Within the scope of the series are monographs, lecture notes, selected contributions from specialized conferences and workshops, as well as selected PhD theses.

Special offer: For all clients with a print standing order we offer free access to the electronic volumes of the Series published in the current year.

Indexed by DBLP, Compendex, EI-Compendex, SCOPUS, Zentralblatt Math, Ulrich's, MathSciNet, Current Mathematical Publications, Mathematical Reviews, MetaPress and Springerlink.

More information about this series at http://www.springer.com/series/5208

Chen Qiu · Jian S. Dai

Analysis and Synthesis of Compliant Parallel Mechanisms—Screw Theory Approach

 Springer

Chen Qiu
Innovation Centre
Nanyang Technological University
Singapore, Singapore

Jian S. Dai
Department of Informatics
King's College London
London, UK

ISSN 1610-7438   ISSN 1610-742X   (electronic)
Springer Tracts in Advanced Robotics
ISBN 978-3-030-48315-9   ISBN 978-3-030-48313-5   (eBook)
https://doi.org/10.1007/978-3-030-48313-5

This Springer imprint is published by the registered company Springer Nature Switzerland AG
The registered company address is: Gewerbestrasse 11, 6330 Cham, Switzerland

The first author would like to dedicate this book to Zhen Li, Wei Qiu and Ye Pan

Foreword

At the dawn of the century's third decade, robotics is reaching an elevated level of maturity and continues to benefit from the advances and innovations in its enabling technologies. These all are contributing to an unprecedented effort to bringing robots to human environment in hospitals and homes, factories and schools; in the field for robots fighting fires, making goods and products, picking fruits and watering the farmland, saving time and lives. Robots today hold the promise for making a considerable impact in a wide range of real-world applications from industrial manufacturing to healthcare, transportation, and exploration of the deep space and sea. Tomorrow, robots will become pervasive and touch upon many aspects of modern life.

The *Springer Tracts in Advanced Robotics (STAR)* is devoted to bringing to the research community the latest advances in the robotics field on the basis of their significance and quality. Through a wide and timely dissemination of critical research developments in robotics, our objective with this series is to promote more exchanges and collaborations among the researchers in the community and contribute to further advancements in this rapidly growing field.

The monograph by Chen Qiu and Jian S. Dai is the outcome of the work accomplished by both authors on screw theory for compliant mechanisms and compliant parallel robots with their presentation, design, stiffness construction, parametrisation and optimisation. Several aspects of mechanics and robotics are mastered, ranging from screw theory and flexible systems to design and optimisation of compliant mechanisms. While conjugating theoretical methods with practical implementation, an excellent way of solving the numerous problems is presented throughout the chapters.

The volume will last long in the field, and future research on the topic will benefit from the impact of the results available with this work. A fine addition to the STAR series!

Naples, Italy   Bruno Siciliano
March 2020   STAR Editor

Preface

As an emerging technology, compliant mechanisms have been used in research and many engineering applications, such as remote centre devices, micro and nano manipulators, continuum manipulators, etc. A compliant mechanism generates a motion based on deformation of flexible elements. As such, to successfully design a compliant mechanism requires a good understanding of both deformation of flexible elements and the combination pattern of them. Deformation evaluation of flexible elements is in the field of solid mechanics, while assembly of flexible elements is more related to a traditional mechanism design. In particular, the latter provides a potential of using the mechanism-equivalence principle for a design of compliant mechanisms.

As a well-established algebra approach in the study of kinematics and dynamics of traditional mechanisms, screw theory has become an increasingly popular tool in design of compliant mechanisms following the fundamental mechanism-equivalence principle. This monograph is the outcome of the work accomplished by both authors on screw theory for compliant mechanisms and compliant parallel robots with their presentation, design, stiffness construction and parametrization and optimization.

The book covers several aspects of mechanics and robotics, ranging from screw theory and flexible systems to design and optimisation of compliant mechanisms and compliant parallel mechanisms. In this book, a unified screw-theory based framework for design of both traditional and compliant mechanisms is proposed, including the description of flexible elements and the stiffness/compliance construction of the whole mechanisms and whole parallel mechanisms. A novel compliant parallel mechanism employing shape-memory-alloy spring based actuators is introduced using a constraint-based approach, and both stiffness analysis and synthesis design problems are tackled. This naturally goes to parameterisation and optimisation with respect to design parameters of ortho-planar springs and leads to development of a novel continuum manipulator with large bending capabilities, and to an original origami-inspired compliant mechanism with good flexibility and controllable motion in a large workspace.

The book presents a comprehensive study on screw theory and its application in the design of compliant mechanisms, with particular focuses on compliant parallel mechanisms. Chapter 2 introduces the theoretical background of screw theory, based on which both compliance characteristics of flexible elements and their integration designs are addressed in Chaps. 3 and 4. This paves a way of analysis and synthesis, where a number of common design topics are covered in Chaps. 5–8.

Chapter 5 looks into synthesis problems of compliant parallel mechanisms at the conceptual-design level, where screw theory is implemented in generating an arrangement of constraint flexible elements according to degrees of freedom of a compliant parallel mechanism. Chapter 6 extends Chap. 5 and addresses the stiffness synthesis problem, where algorithms are developed that are able to synthesize a compliant parallel mechanism not only according to the motion requirement but also according to a stiffness requirement. Chapter 7 looks into the dimensional design issue, where parameterization and optimization analysis of compliant parallel mechanisms are investigated using ortho-planar springs as examples. Finally, Chap. 8 addresses the large deformation of compliant parallel mechanisms using a repelling-screw based approach, where the force equilibrium is established when a compliant parallel mechanism is deformed. In this chapter, Origami-inspired compliant mechanisms are selected to demonstrate the proposed approach.

To help readers better understand each design topic, both simulations and physical experiments are provided in accordance with the mathematical models. The prerequisite for this book is a basic knowledge in linear algebra, kinematics and statics, and solid mechanics. The book is appropriate for researchers, developers, engineers and graduate students with interests in compliant mechanisms and robotics and screw theory.

The book would not have been possible without the help of many people. We would like to thank colleagues in the Group of Advanced Kinematics and Reconfigurable Robotics of King's College London, who offered generous supports in a number of research projects that resulted in the main contents of this book. We acknowledge the support of National Natural Science Foundation of China (NSFC) in the number of 51535008 and the support of Engineering and Physical Science Research Council (EPSRC) in the UK under the grant of EP/E012574/1 and EP/S019790/1. Special thanks to the STAR book series editors Prof. Bruno Siciliano and Prof. Oussama Khatib for valuable suggestions and comments, and Dr. Thomas Ditzinger for the kind help in the publication of this book.

London, UK   Chen Qiu
March 2020   Jian S. Dai

Contents

Acronyms

S	Screw
s	The primary part of a screw
s_0	The second part of a screw
S_{axis}	Screw in Plücker axis coordinate frame
Δ	Screw elliptical polar operator
Ad	Adjoint transformation matrix
$\mathbb{S}$	Screw system
S^r	Reciprocal screw
$\mathbb{S}^r$	Reciprocal screw system
$\mathbb{S}^p$	Repelling screw system
dim	Dimension of a screw system
t	Twist representing an instantaneous velocity or an infinitesimal displacement
ω	The primary part of a twist representing the angular velocity when the twist represents an instantaneous velocity
υ	The second part of a twist representing the linear velocity when the twist represents an instantaneous velocity
θ	The primary part of a twist representing the rotational displacement when the twist represents an infinitesimal displacement
δ	The second part of a twist representing the translational displacement when the twist represents an infinitesimal displacement
T	Twist in Plücker axis coordinate frame
w	Wrench representing a spatial force
f	The primary part of a wrench representing the linear component (pure force)
m	The second part of a wrench representing the angular component (pure moment)
W	Wrench in Plücker axis coordinate frame
$\mathbf{C}$	Compliance matrix
$\mathbf{K}$	Stiffness matrix
k_δ	Stiffness coefficient of a linear spring

c_δ Compliance coefficient of a linear spring
k_θ Stiffness coefficient of a torsional spring
c_θ Compliance coefficient of a torsional spring
$\mathbf{C}_\theta$ Compliance matrix of a revolute joint in the serial configuration
$\mathbf{C}_\delta$ Compliance matrix of a translational joint in the serial configuration
$\mathbf{K}_\theta$ Stiffness matrix of a revolute joint in the parallel configuration
$\mathbf{K}_\delta$ Stiffness matrix of a translational joint in the parallel configuration
$\mathbf{J}_S$ Jacobian matrix
E Young's modulus
G Shear modulus
ν Poisson's ratio

List of Figures

List of Tables

Chapter 1
Introduction

1.1 Modeling of Flexible Elements

Flexible elements have been widely used in the traditional rigid-body mechanisms and robots before they are applied to designing compliant mechanisms. For example, the typical extensional and torsional spring are installed in the translational and revolute joint to provide linear and rotational resistive force. In these applications, they are designed to resist motion in one direction but no other forms of motion. The modelling of the relationship between applied forces and deflections of these simple springs have been well studied, readers can refer to standard mechanical textbooks [2] for their detailed formulations.

Inspired by the application of simple springs in traditional mechanisms, researchers developed flexible elements with similar functions in compliant mechanisms. Successful applications include the beam-type flexures [3, 4], notch hinges [5–7] and emerging crease-type flexures [8–10]. They have relatively simple shapes and can be easily assembled to an integrated platform or be machined in a single-piece material. It is worth noticing that the design of notch hinges and crease-type flexures can be adapted from the beam-type flexures [6], as can be seen from Fig. 1.1.

Further, in modelling flexible elements, various approaches have been proposed with different focuses. Some approaches focus on the initial conceptual design, which usually adopts a skeletal representation of flexible elements, such as the degree-of-freedom. Other approaches pay more attention to the detailed performance of flexible elements, such as the large deformation property.

C. Qiu and J. S. Dai, *Analysis and Synthesis of Compliant Parallel Mechanisms—Screw Theory Approach*, Springer Tracts in Advanced Robotics 139, https://doi.org/10.1007/978-3-030-48313-5_1

Fig. 1.1 The beam-type
flexure and its extensions to
notch hinges and crease-type
flexures, **a** the small-length
beam flexure, **b** the notch
hinge [6], **c**
composite-material type [8],
d multi-layer type [9]

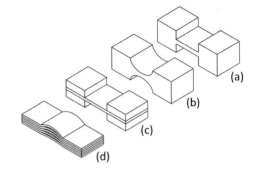

1.1.1 Degree-of-Freedom of Flexible Elements

In early design stages, researchers tend to represent flexible elements in a simple
manner, for example, in forms of degrees of freedom [11]. However, the determina-
tion of the number of degree-of-freedom depends on the categorization of flexure's
deformation under external loads in different directions, which usually begins with
the cantilever-beam deflection modelling under tip loads. Under various load types,
a beam flexure can have three types of deflection, including the axial deformation,
the torsion and the traverse deflections, as can be seen in Fig. 1.2. Particularly, the
traverse deflections include both the translation and rotation in two directions that
are perpendicular to the neutral axis of the beam.

However, the categorization of the degree-of-freedom of a beam flexure is some-
how ambiguous since the magnitude of deflection in different directions is based on
the beam's geometrical shapes and material properties. As a result, it is important
to know the relationship between forces and deflections before classifying flexible
elements directly using the concept of degree-of-freedom. The most widely used
theory is the Euler-Bernoulli beam theory or classical beam theory, which is able to
model the deflections of a beam in a simple fashion. Then Timoshenko beam theory
[12] was proposed by taking into account the shear deformation of beams, making
it possible to describe the deformation of short beams accurately.

The derivation of force/deflection relationship in all directions paves the way for
the further qualitative and quantitative analysis of flexible elements. Qualitative study
of flexible elements is mostly used at the conceptual design stage when designers tend

Fig. 1.2 Deflection types of
a typical beam flexure

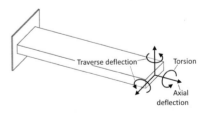

Fig. 1.3 A list of the main degree-of-freedoms of the widely used flexible elements, including **a** beam-type flexure, **b** blade-type flexure, **c** notch-hinge flexure

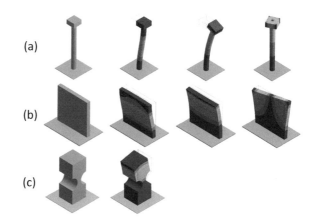

to quickly find out suitable flexible elements that can provide desired mobility rather than detailed force/deflection behaviours. A number of methods [11, 13, 14] have been proposed accordingly, most of which categorize flexible elements according to their main degree-of-freedom that represents the number of their allowed motions. A schematic diagram of the main DOFs of several widely used flexible elements are listed in Fig. 1.3. For a slender-beam type flexure, it is considered to have five degrees of freedom and one constraint along the axial direction. For a blade-type flexure that has comparably large sizes in the length and width direction, it is assumed to have three degrees of freedom, including one translation and one rotation in the traverse direction, and a torsional rotation about the neutral axis. With respect to the notch-hinge flexure (similar to the crease-type flexure), it is considered to have only one rotational degree-of-freedom. Readers can further refer to Chap. 3 for more details of this categorization.

In terms of the quantitative analysis, mostly the compliance/stiffness matrix-based approaches are utilized [15–19]. The idea of using compliance/stiffness matrix is not new, as has been used in the structural matrix analysis [20]. However, the emphasize is different: from the civil and structural engineering discipline, flexible elements are designed for the purpose of building stiffest possible structure; in the design of compliant mechanisms, the priority is to design flexible elements with large flexibility and enough stiffness. In summary, both the qualitative and quantitative approaches provide alternatives to describe the performance of flexible elements, which are proved to be essential for the further integration of flexible elements in designing compliant mechanisms.

1.1.2 Large Deflection of Flexible Elements

Apart from the kinematic description of flexible elements, the large deformation problem has also drawn a substantial amount of attention from researchers. In terms

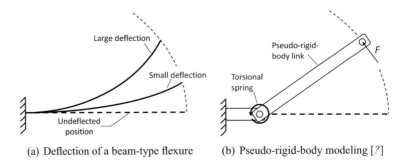

(a) Deflection of a beam-type flexure   (b) Pseudo-rigid-body modeling [?]

Fig. 1.4 Large deformation models of beam-type flexures

of the beam-type flexures, both closed-form solutions and numerical solutions have been proposed to solve this problem for their own advantages. Classic closed-form solutions include elliptic integrals [21, 22] that can give exact formulation to describe the relationship between external loads and beam deflections, but the derivations are complicated and solutions can be found only for relatively simple geometries and loadings. On the other hand, with the rapid development of computational techniques, numerical solutions such as finite element methods (FEM) [23, 24] have been largely used gradually. The fundamental idea is to divide a flexure into a finite number of small elements. By analyzing and integrating the load/deflection characteristics of small elements, the overall performance of the whole flexure can be evaluated accordingly. FEM based approaches are suitable when the shape of a flexible element is complex but at the cost of computational efficiency and accuracy. Generally, FEM based approaches are mostly used at a later performance-validation stage for the developed compliant mechanisms (Fig. 1.4).

Another useful method for modelling large deformation of planar beams is the pseudo-rigid-body model (PRBM) approach [1]. In the PRBM approach, a flexible beam is modeled with two rigid links, one link represents the rigid segment and the other one represents the motion of the compliant segment and is called the pseudo-rigid-body link. As a result, the tip motion of the slender beam can be simulated by the rotational motion of the pseudo-rigid-link, and the flexible beam's resistance to external load can be modeled by adding a torsional spring at the pseudo joint. For the formulation simplicity, PRBM approach is effective and easily accessible at the initial design stage of compliant platforms where large beam deformation occurs. Also, pseudo-rigid-body models make it possible to model flexible bodies as rigid bodies, thus allowing the application of analysis and synthesis methods from rigid-body mechanisms [25–27].

In terms of flexible elements such as notch-hinges and crease-type flexures, they have been designed mainly to provide one rotational degree-of-freedom, making it possible to model them directly as revolute joints with embedded rotational stiffness, which becomes the most important feature in describing the performance of these

Fig. 1.5 Large deformation of single-DOF flexible elements, including: **a** notch-hinge flexure, **b** crease-type flexure

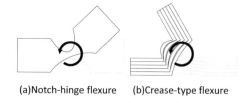

(a)Notch-hinge flexure   (b)Crease-type flexure

single degree-of-freedom flexible elements. Figure 1.5 presents two typical single DOF flexible elements, including a notch-hinge flexure and a crease-type flexure. Notch-hinge flexures are widely utilized in designing flexure-based precision machine [17, 28], both closed-form solutions [29] and FEM-based approaches [7] have been proposed to explore the rotational stiffness of notch hinges. In terms of crease-type flexures, they are frequently appeared in the recently developed foldable and printable robots [30, 31]. Due to the complex multi-layer structure, mostly experiments [32, 33] were conducted to obtain the rotational stiffness of crease-type flexures.

1.2 Integration of Flexible Elements

Following the deflection/compliance modelling of flexible elements, proper integration of them becomes important for the good functionality of designed compliant mechanisms. As discussed in Sect. 1.1, flexible elements have been designed to have similar functions as traditional rigid joints, thus it is natural to design compliant mechanisms by treating them as equivalent traditional mechanisms. However, different flexible elements have different spatial compliance performances. As shown in Fig. 1.2, a beam-type or blade-type flexure has equal or more degrees of freedom than degrees of constraint, while a notch-hinge or crease-type flexure has only one degree of freedom. On the other hand, most rigidly connected joints have fewer degrees of freedom compared to constraints in traditional mechanisms. This indicates the necessity of adopting different approaches in the integration of flexible elements. Two types of design approaches have been proposed to address this issue, namely the constraint-based design approach and mechanism-equivalent approach. In the constraint-based design approach, flexible elements are treated as constraints, which is particularly useful for flexures such as the beam-type flexures. On the other hand, the mechanism-equivalent approach also treats a compliant mechanism as a traditional mechanism but directly uses traditional joints to replace flexible elements. As a result, the kinematics and statics of a compliant mechanism can be addressed properly by using the mechanism-equivalent method. In the following sections, both constraint-based design approach and mechanism-equivalent method are discussed in detail.

1.2.1 Constraint-Based Design Approach

Constraint-based approach was first introduced in [11, 15]. In [11], the author explored the mechanical connection types between objects, which inspired the proposal of the constraint-line concept that can visually describe the positions and orientations of constraints in space. The nature of flexible elements that constrain motions in certain directions makes the constraint-line pattern an ideal tool to represent flexible elements, particularly beam flexures which mainly resist motion along their axes. Based on the constraint-line representation, the total number of constraints exerted on the functional platform can be obtained, which can be further utilized to determine the allowed degrees of freedom of developed compliant mechanisms. One successful example is the planar XY-motion flexure platform based on beam-type flexures [4], which is shown in Fig. 1.6a.

However, the initial constraint-based design approach heavily relies on the researcher's experience to properly assemble flexible elements for given motions, which lacks rigorous mathematical description and the result cannot be easily generalized. Hopkins and Culpepper [34–36] further improved the constraint-based approach by introducing mathematical tools to analytically describe the constraints introduced by flexible elements. In their proposed freedom and constraint topology approach, the freedom and constraint space of a given compliant mechanism were described in line patterns, particularly screw theory [37] was used for the first time to interpret the line patterns and establish the relationship between freedom and constraint space visually.

Further, Su and Tari [19] revealed the constraint-based design approach utilize one special type of screw systems named the line screw system, and identified the relationship between the freedom space and constraint space as *reciprocal* [38] in the framework of screw theory. This makes it possible to synthesize the design of a compliant mechanism using constraints according to its required degrees of freedom analytically. Further algorithms for synthesizing the actuation spaces were also proposed recently [36, 39, 40] that can determine the location and orientation of actuators for given motions of compliant mechanisms. A compliant parallel platform was designed accordingly using the developed algorithms, as can be seen in Fig. 1.6b.

1.2.2 Mechanism-Equivalent Approach

The key to implement constraint-based design is to treat flexible elements as constraints, which is particularly useful when the used flexible elements have fewer constraints than degrees of freedom. However, in terms of flexible elements such as notch hinges or flexure pivots that mainly provide one degree of freedom, the direct mechanism-equivalent approach can be more efficient to solve the design problems.

The utility of mechanism-equivalent approach follows the popular application of flexure parallel mechanism for precise micro-manipulation early in 2000 and then

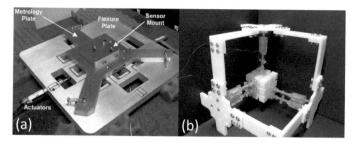

Fig. 1.6 Compliant mechanisms built using the constriant-based design approach, including **a** a planar 2DOF parallel platform [4], **b** a spatial shape-memory-alloy actuated platform [41]

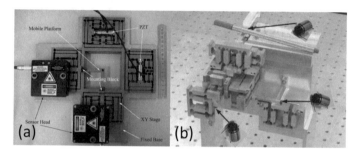

Fig. 1.7 Typical flexure parallel platforms using mechanism-equivalent approach, including **a** a planar 2DOF platform [18], **b** a spatial 3DOF platform [47]

attracts continuous attention [16–18, 28, 42–46]. Figure 1.7 presents two typical types of flexure parallel platforms, including a planar flexure parallel platform for decoupled XY-motion realization [18] in Fig. 1.7a, and a spatial flexure platform that can realize decoupled XYZ motions [47] in Fig. 1.7b. In these designs, mainly notch-hinge type flexures were utilized to build up platforms for the ease of integrated manufacturing. Since a notch hinge can be modelled as a revolute joint with inherent torsional stiffness, a flexure parallel platform can be treated as a parallel mechanism with revolute joints. As a consequence, the force/deflection characteristics of flexure parallel platforms were addressed by conducting the kinematics and statics analysis of the equivalent parallel mechanisms [42, 44–46]. Also the force/deflection performance were investigated by developing the corresponding stiffness/compliance matrix of equivalent mechanisms [16–18, 28, 43].

Apart from notch-hinge enabled parallel platforms, recently more light-weight, smart-material enabled foldable structures and robots have been developed [31, 48–51], as can be seen from Fig. 1.8. These designs utilize the origami art to fold a flat-sheet material into a 3D structure which includes crease-type flexures and panels that connect creases. Due to the fact mainly the crease flexures generate rotational motions, a compliant origami structure can also be modelled using the mechanism-equivalent approach [52] by treating creases as revolute joints and panels as links. This immediately allows the foldability (kinematics) analysis of compliant origami

Fig. 1.8 Origami-inspired
compliant structures, **a** a
deformable wheel [50], **b** a
3DOF compliant structure
[66]

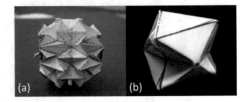

structures as well as novel designs such as palm-foldable hand [53] and innovative
mechanisms [54–58]. Further, the compliance performance of origami compliant
structures is also evaluated in the same framework. Related research outcome include
the folding-force evaluation of single origami creases [30, 32, 59, 60] as well as the
whole origami structures [9, 33, 61–64]. Recently a repelling-screw based force
analysis approach [65] was proposed to conduct force and compliance analysis of
origami structures.

1.3 Screw-Theory Based Approach

As can be seen from Sect. 1.2, the integration of flexible elements are highly
related to the modelling of flexible elements. Two main approaches have been
discussed, including the constraint-based design approach and the mechanism-
equivalent approach. They are intrinsically the same but are different in the way
of modelling flexible elements: constraint-based design approach treats them as constraints, while mechanism-equivalent approach treats them as degrees of freedom.

It is worth noticing screw theory has become an increasingly popular tool in
designing and modelling compliant mechanisms, as has been used in both the
constraint-based design [19, 34] and the mechanism-equivalent design [9, 33]. This is
not a coincidence but a natural reflection of the fundamental idea of using rigid-body
replacement principle to design compliant mechanisms.

1.3.1 A Brief History of Screw Theory

Screw theory is an algebra for describing forces and motions, that appear in the
kinematics and dynamics of rigid bodies, in dual-vector forms. Chasles and Poinsots'
geometrical expositions of wrenches and twists are the two fundamental theorems of
screw theory. Michel Chasles proved that a rigid motion in three dimensions can be
represented by a rotation about an axis followed by a translation along this same axis.
This motion is generally referred as a *screw motion*. The infinitesimal version of a
screw motion is called a *twist*, it can be used to describe the instantaneous velocity
or displacement of a rigid body in terms of its linear and angular component. In duel

to the rigid body motion, Louis Poinsot discovered that any system of forces acting on a rigid body can be replaced by a single force applied along an axis combined with a torque about the same axis. This force and torque combination is called a *wrench*. Later in 1900, Sir Robert Stawell Ball established the mathematical framework of screw theory that unifies the form of twist and wrench for application in kinematics and statics of mechanisms [37].

Twist and wrench are the two fundamental concepts in screw theory, and they are naturally related to each other through the stiffness/compliance of a mechanism or robotic system. The study of the compliance property can be dated back to Ball's treatise [37] which formulated the compliance relationship as an eigenvalue problem. Dimentberg [67] used screw theory to study the static and small vibrations of an elastically suspended rigid body, where the relationship between displacement screws and wrenches was investigated in a form of the stiffness matrix. Further Loncaric [68] analyzed the stiffness matrix using Lie groups. He first showed that only stiffness matrix with zero trace off-diagonal matrices are realizable by line springs. This leads to the subsequent investigation of invariance properties of robotic compliance using screw theory [69–72], as well as stiffness decomposition problems [73–75]. In parallel, screw theory has also become a popular tool to construct the Jacobian matrix of both serial and parallel manipulators [76, 77], which paves the way for further stiffness/compliance analysis of them [78, 79]. For example, the stiffness properties of a number of parallel manipulators have been evaluated [80–82] by developing their stiffness matrices using screw theory.

1.3.2 Compliant-Mechanism Design Using Screw Theory

The application of screw theory in designing compliant mechanisms follows the fundamental mechanism-equivalence design principle, which can be categorized into two levels: conceptual-design level and dimensional-design level. On the conceptual-design level, screw theory bridges the visualization of compliant-mechanism design and the rigorous mathematical interpretation. Visually, screw theory describes both the freedom and constraint in line patterns in space, thus the motion (constraint) of a flexible element and the integrated compliant mechanism can be well represented in this framework. More importantly, the mathematical formulation of twists and constraints, as well as their relationships makes it possible to design compliant mechanisms either by picking one configuration from a comprehensive categorization [34] according to design requirements, or by generating the layout of constraints and actuators using algorithms automatically [19, 40].

On the dimensional-design level, screw theory can be used to describe the force/deflection characteristics of flexible elements, as well as that of compliant mechanisms in forms of compliance/stiffness matrix. Successful examples include the compliance modelling of flexible elements such as slender beams and coil springs [83, 84], as well as compliant mechanisms such as a remote-center device [85], a vibratory bowl feeder [86] and a flexure parallel platform for micromanipulation [28].

In addition, screw theory makes it possible to evaluate the invariance properties of built compliant mechanisms through the developed compliance/stiffness matrix, such as eigenscrew evaluation [72], and further decompose compliance/stiffness using the obtained invariance properties [74, 75].

1.4 Book Overview

Though screw theory has been an established approach for designing and analyzing traditional mechanisms and robots, its application in compliant mechanisms has a relatively short history. As a result, there is a need to look into the theoretical foundation of screw theory and explores the advantages and limitations of proposed screw-theory based approaches.

To address this issue, this book goes through screw theory and tries to develop a systematic approach to design compliant mechanisms in the framework of screw theory, with a particular focus on the compliant parallel mechanisms. Based on the establishment of the design framework, we can address various design problems that appear in compliant parallel mechanisms by solving the equivalent design problems in traditional mechanisms, such as the conceptual-design and dimensional-design problem, the mechanism analysis and synthesis problem, as well as the large deformation problem. Further several practical compliant parallel mechanisms and their equivalents are given to demonstrate the utility of the addressed compliant-mechanism design problems.

1.4.1 Motivation

The primary motivates of this book can be listed as follows:

1. Explore screw theory and establish its theoretical foundation in designing compliant mechanisms. This can be achieved by following the mechanism-equivalence design principle, and unifying the analysis and design of both traditional mechanisms and compliant mechanisms in the same framework of screw theory, with particular focuses on the compliance study of flexible elements as well as the integration of them.
2. Explore the synthesis design problems of compliant parallel mechanisms at both the conceptual-design level and the dimensional-design level. At the conceptual-design level where mainly degrees of freedom and constraints are considered, screw theory will be implemented to develop algorithms that can automatically generate the arrangement of constraints as well as actuators for given motions of compliant mechanisms. At the dimensional-design level where the stiffness of compliant parallel mechanisms is considered, algorithms will also be developed that is able to decompose the pre-determined stiffness matrix of a compliant paral-

lel mechanism into a configuration of flexible elements, thus achieving complete synthesis without knowing the layout of the compliant mechanism.

3. Explore the optimization design of compliant parallel mechanisms. Compared to the stiffness synthesis of compliant parallel mechanisms, optimization design aims to optimize the dimensional parameters of a compliant mechanism instead of creating the layout of the structure, which is particularly useful when the configuration of compliant mechanisms have already been determined according to the application constraints.

4. Explore the large deformation property of compliant parallel mechanisms. When a compliant parallel mechanism undertakes an external load and generates a relatively large deformation, alternative approaches that can conduct force/deflection analysis should be explored instead of the static compliance/stiffness matrix based approaches.

1.4.2 Organization of Remaining Chapters

In accordance with this, the remainder of this book is organized as follows. The theoretical foundation of screw theory is established from Chaps. 2 to 3. Chapter 2 presents a comprehensive study of screw theory and introduces the two most important concepts, namely twist and wrench, to describe the force and deflection of both flexible elements and compliant mechanisms. Then Chap. 3 investigates flexible elements in terms of their degrees of freedom and compliance/stiffness performance, and Chap. 4 further looks into the integration of flexible elements following the design of traditional mechanisms.

Based on the developed design framework, a number of compliant mechanism design problems are proposed and solved from Chaps. 5 to 8. Chapter 5 looks into the conceptual design problem and proposes a constraint-based design approach [87] in the framework of screw theory, which can automatically synthesize the layout of constraints and actuators for given degrees-of-freedom of compliant mechanisms [41]. Chapter 6 extends the synthesis design to the stiffness of compliant mechanisms. In this chapter, both the stiffness analysis and synthesis are explored using the invariant properties of compliant mechanisms. Particularly a matrix-partition approach is proposed [88] that is able to decompose the stiffness matrix of a compliant parallel mechanism using selected constraint flexures.

Chapter 7 focuses on solving the optimization design problem of compliant mechanisms. Ortho-planar spring is selected as an example, where its six-dimensional compliance performance is evaluated in forms of compliance matrices and further optimized by conducting parameterization and optimization analysis of design parameters [89]. Chapter 8 further investigates the large deformation problem of compliant mechanisms using origami-inspired compliant mechanisms. Experimental tests are conducted to explore the stiffness behaviours of crease-type flexures [90], based on which the force transmission of compliant origami structures undertaking large deformations are evaluated using the mechanism equivalent approach.

Particularly, a novel repelling-screw based approach is developed for the first time that is able to conduct forward force analysis of compliant origami mechanisms [65]. Notably, a series of continuum manipulators and bio-inspired robots are further developed using the evaluated compliant origami modules [31, 66].

References

1. Howell, L.L.: Compliant Mechanisms. Wiley-Interscience (2001)
2. Shigley, J.E., Mischke, C.R., Budynas, R.G., Liu, X., Gao, Z.: Mechanical Engineering Design. vol. 89. McGraw-Hill, New York (1989)
3. Parise, J.J., Howell, L.L., Magleby, S.P.: Ortho-planar linear-motion springs. Mech. Mach. Theory **36**(11), 1281–1299 (2001)
4. Awtar, S., Slocum, A.H.: Constraint-based design of parallel kinematic XY flexure mechanisms. J. Mech. Des. **129**(8), 816–830 (2007)
5. Goldfarb, M., Speich, J.E.: A well-behaved revolute flexure joint for compliant mechanism design. J. Mech. Des. **121**(3), 424–429 (1999)
6. Smith, S.T.: Flexures: Elements of Elastic Mechanisms. CRC Press (2000)
7. Zhang, S., Fasse, E.D.: A finite-element-based method to determine the spatial stiffness properties of a notch hinge. J. Mech. Des. **123**(1), 141–147 (2001)
8. Wood, R., Avadhanula, S., Sahai, R., Steltz, E., Fearing, R.: Microrobot design using fiber reinforced composites. J. Mech. Des. **130**(5), 052304 (2008)
9. Dai, J.S., Cannella, F.: Stiffness characteristics of carton folds for packaging. J. Mech. Des. **130**(2) (2008)
10. McGough, K., Ahmed, S., Frecker, M., Ounaies, Z.: Finite element analysis and validation of dielectric elastomer actuators used for active origami. Smart Mater. Struct. **23**(9), 094002 (2014)
11. Blanding, D.L.: Exact constraint: machine design using kinematic processing. American Society of Mechanical Engineers (1999)
12. Timoshenko, S.P., Goodier, J.: Theory of elasticity. Int. J. Bulk Solids Storage Silos **1**(4) (2014)
13. Su, H., Yue, C.: Type synthesis of freedom and constraint elements for design of flexure mechanisms. J. Mech. Sci **4**(2), 263–277 (2013)
14. Zhang, Y., Su, H.-J., Liao, Q.: Mobility criteria of compliant mechanisms based on decomposition of compliance matrices. Mech. Mach. Theory **79**, 80–93 (2014)
15. Hale, L.C.: Principles and techniques for designing precision machines. Technical report, Lawrence Livermore National Lab., CA, USA (1999)
16. Koseki, Y., Tanikawa, T., Koyachi, N., Arai, T.: Kinematic analysis of a translational 3-DOF micro-parallel mechanism using the matrix method. Adv. Robot. **16**(3), 251–264 (2002)
17. Pham, H.-H., Chen, I.-M.: Stiffness modeling of flexure parallel mechanism. Precis. Eng. **29**(4), 467–478 (2005)
18. Li, Y., Xu, Q.: Design and analysis of a totally decoupled flexure-based XY parallel micromanipulator. IEEE Trans. Robot. **25**(3), 645–657 (2009)
19. Su, H.-J., Tari, H.: Realizing orthogonal motions with wire flexures connected in parallel. J. Mech. Des. **132**, 121002 (2010)
20. Pilkey, W.D.: Formulas for Stress, Strain, and Structural Matrices. Wiley (1993)
21. Bisshopp, K., Drucker, D.: Large deflection of cantilever beams. Q. Appl. Math. **3**(1) (1945)
22. Frisch-Fay, R.: Flexible Bars. Butterworths (1962)
23. Zienkiewicz, O.C., Taylor, R.L.: The Finite Element Method: Solid Mechanics, vol. 2. Butterworth-Heinemann (2000)
24. De Borst, R., Crisfield, M.A., Remmers, J.J., Verhoosel, C.V.: Nonlinear Finite Element Analysis of Solids and Structures. Wiley (2012)

25. Su, H.-J.: A pseudorigid-body 3R model for determining large deflection of cantilever beams subject to tip loads. J. Mech. Robot. **1**(2), 021008 (2009)
26. Xu, P., Jingjun, Y., Guanghua, Z., Shusheng, B., Zhiwei, Y.: Analysis of rotational precision for an isosceles-trapezoidal flexural pivot. J. Mech. Des. **130**(5), 052302 (2008)
27. Xu, P., Jingjun, Y., Guanghua, Z., Shusheng, B.: The stiffness model of leaf-type isosceles-trapezoidal flexural pivots. J. Mech. Des. **130**(8), 082303 (2008)
28. Pham, H.-H., Yeh, H.-C., Chen, I.-M.: Micromanipulation system design based on selective actuation mechanisms. Int. J. Robot. Res. **25**(2), 171–186 (2006)
29. Lobontiu, N., Garcia, E.: Two-axis flexure hinges with axially-collocated and symmetric notches. Comput. Struct. **81**(13), 1329–1341 (2003)
30. Felton, S.M., Tolley, M.T., Shin, B., Onal, C.D., Demaine, E.D., Rus, D., Wood, R.J.: Self-folding with shape memory composites. Soft Matter **9**(32), 7688–7694 (2013)
31. Zhang, K., Qiu, C., Dai, J.S.: Helical Kirigami-enabled centimeter-scale worm robot with shape-memory-alloy linear actuators. J. Mech. Robot. **7**(2), 021014 (2015)
32. Beex, L., Peerlings, R.: An experimental and computational study of laminated paperboard creasing and folding. Int. J. Solids Struct. **46**(24), 4192–4207 (2009)
33. Qiu, C., Aminzadeh, V., Dai, J.S.: Kinematic analysis and stiffness validation of origami cartons. J. Mech. Des. **135**(11), 111004 (2013)
34. Hopkins, J.B.: Design of flexure-based motion stages for mechatronic systems via freedom, actuation and constraint topologies (FACT). Ph.D. thesis, Massachusetts Institute of Technology (2010)
35. Hopkins, J.B., Culpepper, M.L.: Synthesis of multi-degree of freedom, parallel flexure system concepts via freedom and constraint topology (fact)-part I: Principles. Precis. Eng. **34**(2), 259–270 (2010)
36. Hopkins, J.B., Culpepper, M.L.: A screw theory basis for quantitative and graphical design tools that define layout of actuators to minimize parasitic errors in parallel flexure systems. Precis. Eng. **34**(4), 767–776 (2010)
37. Ball, R.S.: A Treatise on the Theory of Screws. Cambridge University Press (1900)
38. Dai, J.S., Rees Jones, J.: Interrelationship between screw systems and corresponding reciprocal systems and applications. Mech. Mach. Theory **36**(5), 633–651 (2001)
39. Qiu, C., Yu, J., Li, S., Su, H.-J., Zeng, Y.: Synthesis of actuation spaces of multi-axis parallel flexure mechanisms based on screw theory. In: ASME 2011 International Design Engineering Technical Conferences and Computers and Information in Engineering Conference, pp. 181–190. American Society of Mechanical Engineers (2011)
40. Yu, J.J., Li, S.Z., Qiu, C.: An analytical approach for synthesizing line actuation spaces of parallel flexure mechanisms. J. Mech. Des. **135**(12), 124501–124501 (2013)
41. Qiu, C., Zhang, K.T., Dai, J.S.: Constraint-based design and analysis of a compliant parallel mechanism using SMA-spring actuators. In: Proceedings of ASME 2014 International Design Engineering Technical Conferences and Computers and Information in Engineering Conference, New York, Buffalo, USA, August 17–20, 2014. ASME, New York (2014)
42. Yi, B.-J., Chung, G.B., Na, H.Y., Kim, W.K., Suh, I.H.: Design and experiment of a 3-DOF parallel micromechanism utilizing flexure hinges. IEEE Trans. Robot. Autom. **19**(4), 604–612 (2003)
43. Kang, B.H., Wen, J.T.-Y., Dagalakis, N., Gorman, J.J.: Analysis and design of parallel mechanisms with flexure joints. IEEE Trans. Robot. **21**(6), 1179–1185 (2005)
44. Li, Y., Xu, Q.: A novel design and analysis of a 2-DOF compliant parallel micromanipulator for nanomanipulation. IEEE Trans. Autom. Sci. Eng. **3**(3), 247–254 (2006)
45. Yao, Q., Dong, J., Ferreira, P.M.: Design, analysis, fabrication and testing of a parallel-kinematic micropositioning XY stage. Int. J. Mach. Tools Manuf. **47**(6), 946–961 (2007)
46. Li, Y., Xu, Q.: A totally decoupled piezo-driven XYZ flexure parallel micropositioning stage for micro/nanomanipulation. IEEE Trans. Autom. Sci. Eng. **8**(2), 265–279 (2011)
47. Tang, X., Chen, I., et al.: A large-displacement and decoupled XYZ flexure parallel mechanism for micromanipulation. In: 2006 IEEE International Conference on Automation Science and Engineering, CASE'06, pp. 75–80. IEEE (2006)

48. Carroll, D.W., Magleby, S.P., Howell, L.L., Todd, R.H., Lusk, C.P.: Simplified manufacturing through a metamorphic process for compliant ortho-planar mechanisms. In: ASME 2005 International Mechanical Engineering Congress and Exposition, pp. 389–399. American Society of Mechanical Engineers (2005)
49. Winder, B.G., Magleby, S.P., Howell, L.L.: Kinematic representations of pop-up paper mechanisms. J. Mech. Robot. 1(2), 021009 (2009)
50. Lee, D.-Y., Kim, J.-S., Kim, S.-R., Koh, J.-S., Cho, K.-J.: The deformable wheel robot using magic-ball origami structure. In: Proceedings of the 2013 ASME Design Engineering Technical Conference, Portland, OR (2013)
51. Vander Hoff, E., Jeong, D., Lee, K.: Origamibot-I: a thread-actuated origami robot for manipulation and locomotion. In: 2014 IEEE/RSJ International Conference on Intelligent Robots and Systems (IROS 2014), pp. 1421–1426. IEEE (2014)
52. Dai, J.S., Jones, J.R.: Mobility in metamorphic mechanisms of foldable/erectable kinds. J. Mech. Des. 121(3), 375–382 (1999)
53. Dai, J.S., Wang, D., Cui, L.: Orientation and workspace analysis of the multifingered metamorphic hand–metahand. IEEE Trans. Robot. 25(4), 942–947 (2009)
54. Wilding, S.E., Howell, L.L., Magleby, S.P.: Spherical lamina emergent mechanisms. Mech. Mach. Theory 49, 187–197 (2012)
55. Bowen, L., Frecker, M., Simpson, T.W., von Lockette, P.: A dynamic model of magneto-active elastomer actuation of the waterbomb base. In: ASME 2014 International Design Engineering Technical Conferences and Computers and Information in Engineering Conference, pp. V05BT08A051–V05BT08A051. American Society of Mechanical Engineers (2014)
56. Yao, W., Dai, J.S.: Dexterous manipulation of origami cartons with robotic fingers based on the interactive configuration space. J. Mech. Des. 130(2), 022303 (2008)
57. Zhang, K., Fang, Y., Fang, H., Dai, J.S.: Geometry and constraint analysis of the three-spherical kinematic chain based parallel mechanism. J. Mech. Robot. 2(3), 031014 (2010)
58. Wei, G., Dai, J.S.: Origami-inspired integrated planar-spherical overconstrained mechanisms. J. Mech. Des. 136(5), 051003 (2014)
59. Ahmed, S., Lauff, C., Crivaro, A., McGough, K., Sheridan, R., Frecker, M., von Lockette, P., Ounaies, Z., Simpson, T., Lien, J., et al.: Multi-field responsive origami structures: preliminary modeling and experiments. ASME Paper No. DETC2013-12405 (2013)
60. Delimont, I.L., Magleby, S.P., Howell, L.L.: Evaluating compliant hinge geometries for origami-inspired mechanisms. J. Mech. Robot. 7(1), 011009 (2015)
61. Mentrasti, L., Cannella, F., Pupilli, M., Dai, J.S.: Large bending behavior of creased paperboard. I. Experimental investigations. Int. J. Solids Struct. 50(20), 3089–3096 (2013)
62. Mentrasti, L., Cannella, F., Pupilli, M., Dai, J.S.: Large bending behavior of creased paperboard. II. Structural analysis. Int. J. Solids Struct. 50(20), 3097–3105 (2013)
63. Hanna, B.H., Lund, J.M., Lang, R.J., Magleby, S.P., Howell, L.L.: Waterbomb base: a symmetric single-vertex bistable origami mechanism. Smart Mater. Struct. 23(9), 094009 (2014)
64. Hanna, B.H., Magleby, S., Lang, R.J., Howell, L.L.: Force-deflection modeling for generalized origami waterbomb-base mechanisms. J. Appl. Mech. (2015)
65. Qiu, C., Zhang, K., Dai, J.S.: Repelling-screw based force analysis of origami mechanisms. J. Mech. Robot. 15(1122), 1 (2015)
66. Zhang, K., Qiu, C., Dai, J.S.: An extensible continuum robot with integrated origami parallel modules. J. Mech. Robot. (2015)
67. Dimentberg, F.M.: The screw calculus and its applications in mechanics. Technical report, DTIC Document (1968)
68. Loncaric, J.: Normal forms of stiffness and compliance matrices. IEEE J. Robot. Autom. 3(6), 567–572 (1987)
69. Lipkin, H., Patterson, T.: Geometrical properties of modelled robot elasticity: Part I-decomposition. In: 1992 ASME Design Technical Conference, Scottsdale, DE, vol. 45, pp. 179–185 (1992)
70. Lipkin, H., Patterson, T.: Geometrical properties of modelled robot elasticity: Part II–center of elasticity. In: 22nd Biennal Mechanisms Conference Robotics, Spatial Mechanisms, and Mechanical Systems: ASME Design Technical Conferences, pp. 13–16. Scottsdale, AZ (1992)

71. Patterson, T., Lipkin, H.: Structure of robot compliance. Trans. Am. Soc. Mech. Eng. J. Mech. Des. **115**, 576–576 (1993)
72. Patterson, T., Lipkin, H.: A classification of robot compliance. Trans. Am. Soc. Mech. Eng. J. Mech. Des. **115**, 581–581 (1993)
73. Ciblak, N., Lipkin, H.: Synthesis of cartesian stiffness for robotic applications. In: 1999 Proceedings of the IEEE International Conference on Robotics and Automation, vol. 3, pp. 2147–2152. IEEE (1999)
74. Huang, S., Schimmels, J.M.: The bounds and realization of spatial stiffnesses achieved with simple springs connected in parallel. IEEE Trans. Robot. Autom. **14**(3), 466–475 (1998)
75. Huang, S., Schimmels, J.M.: The eigenscrew decomposition of spatial stiffness matrices. IEEE Trans. Robot. Autom. **16**(2), 146–156 (2000)
76. Waldron, K., Wang, S.-L., Bolin, S.: A study of the Jacobian matrix of serial manipulators. J. Mech. Des. **107**(2), 230–237 (1985)
77. Joshi, S.A., Tsai, L.-W.: Jacobian analysis of limited-DOF parallel manipulators. In: ASME 2002 International Design Engineering Technical Conferences and Computers and Information in Engineering Conference, pp. 341–348. American Society of Mechanical Engineers (2002)
78. Tsai, L.-W.: Robot Analysis: The Mechanics of Serial and Parallel Manipulators. Wiley (1999)
79. Huang, T., Zhao, X., Whitehouse, D.J.: Stiffness estimation of a tripod-based parallel kinematic machine. IEEE Trans. Robot. Autom. **18**(1), 50–58 (2002)
80. Li, Y., Xu, Q.: Stiffness analysis for a 3-PUU parallel kinematic machine. Mech. Mach. Theory **43**(2), 186–200 (2008)
81. Xu, Q., Li, Y.: An investigation on mobility and stiffness of a 3-DOF translational parallel manipulator via screw theory. Robot. Comput. Integr. Manuf. **24**(3), 402–414 (2008)
82. Kim, H.S., Lipkin, H.: Stiffness of parallel manipulators with serially connected legs. J. Mech. Robot. **6**(3), 031001 (2014)
83. Selig, J., Ding, X.: A screw theory of static beams. In: 2001 Proceedings of the IEEE/RSJ International Conference on Intelligent Robots and Systems, vol. 1, pp. 312–317. IEEE (2001)
84. Ding, X., Selig, J.M.: On the compliance of coiled springs. Int. J. Mech. Sci. **46**(5), 703–727 (2004)
85. Ciblak, N., Lipkin, H.: Design and analysis of remote center of compliance structures. J. Robot. Syst. **20**(8), 415–427 (2003)
86. Dai, J.S., Xilun, D.: Compliance analysis of a three-legged rigidly-connected platform device. J. Mech. Des. **128**(4), 755–764 (2006)
87. Yu, J.J., Li, S.Z., Qiu, C.: An analytical approach for synthesizing line actuation spaces of parallel flexure mechanisms. J. Mech. Des. **135**(12), 124501 (2013)
88. Qiu, C., Dai, J.S.: Constraint stiffness construction and decomposition of a SPS orthogonal parallel mechanism. In: Proceedings of ASME 2015 International Design Engineering Technical Conferences and Computers and Information in Engineering Conference, Massachusetts, Boston, USA, August 2–5, 2015. ASME (2015)
89. Qiu, C., Qi, P., Liu, H., Althoefer, K., Dai, J.S.: Six-dimensional compliance analysis and validation of ortho-planar springs. J. Mech. Des. (2016)
90. Qiu, C., Vahid, A., Dai, J.S.: Kinematic analysis and stiffness validation of origami cartons. J. Mech. Des. **135**(11), 111004 (2013)

Chapter 2
An Introduction to Screw Theory

2.1 Introduction

As a well established algebraic approach, screw theory has been widely utilized to analyze traditional mechanisms. Screw theory is an algebra for describing forces and motions that appear in the kinematics of rigid bodies. Michel Chasles (1830) proved that a rigid motion in three dimensions can be represented by a rotation about an axis followed by a translation along this same axis. This motion is generally referred as a *screw motion*. The infinitesimal version of a screw motion is called a *twist*, it can be used to describe the instantaneous velocity or displacement of a rigid body. In duel to the rigid body motion, Louis Poinsot (1806) discovered that any system of forces acting on a rigid body can be replaced by a single force applied along an axis combined with a torque about the same axis. This force and torque combination is called a *wrench*. Later in 1900, Sir Robert Stawell Ball [1] established the mathematical framework of screw theory that unifies the twist and wrench in the form of a screw for application in kinematics and statics of mechanisms.

An aggregate of all screws forms a screw system which is a linear span of a set of linearly independent screws. The interest in screw systems and their relationships dates back to early studies by Ball [1] and was further investigated by Dimentberg [2] and Hunt [3] and many other researchers. The most widely evaluated relationship is the *reciprocal* relationship [4], which has been shown to be essential in the motion and force analysis of both serial and parallel mechanisms [5]. Apart from the reciprocal screw system, Ohwovoriole and Roth [6] proposed the repelling screw system in completion of the reciprocal screw system and used it to solve contact problems.

Further, the concepts of the screw and screw system are extended to design compliant mechanisms. For instance, one successful application is the constraint-based approach [7] which is interpreted using screw theory. The fundamental idea is to represent flexible elements as constraint wrenches, thus the required motion of a com-

© The Editor(s) (if applicable) and The Author(s), under exclusive license to Springer Nature Switzerland AG 2021
C. Qiu and J. S. Dai, *Analysis and Synthesis of Compliant Parallel Mechanisms—Screw Theory Approach*, Springer Tracts in Advanced Robotics 139, https://doi.org/10.1007/978-3-030-48313-5_2

Fig. 2.1 The direction and
position vector of a screw

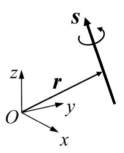

pliant mechanism can be used to generate the layout of flexible elements according
to the reciprocal relationship. This provides the ground for the conceptual design of
compliant mechanisms and leads to several successful applications [8–10].

2.2   Definition of a Screw and the Screw Operations

A screw [11] is a geometrical entity which can be described as a line vector with
a pitch that is a scalar coefficient of the secondary part of the line vector. It can be
written as a six-dimensional vector with the following definition as

$$S = \begin{bmatrix} s \\ s_0 \end{bmatrix} = \begin{bmatrix} s \\ r \times s + hs \end{bmatrix} \tag{2.1}$$

which consists of two parts, the primary part s represents a direction vector in a
spatial space, and the second part s_0 is the sum of $r \times s$ and hs. In a three-dimensional
Euclidean space, both s and s_0 are 3×1 vectors and together S is a 6×1 vector.
r is the position vector that determines the location of s, while h is the pitch of s
which represents the ratio of magnitude of hs with respect to that of s. A geometrical
interpretation of a screw is shown in Fig. 2.1.

From Eq. (2.1) we are able to calculate both the pitch h and position vector r
using both scalar product and cross product of vectors once we know the vector form
of S. For instance, we can use a scalar product to calculate the pitch h as

$$h = \frac{s^\mathsf{T} s_0}{\|s\|^2} \tag{2.2}$$

Similarly, the position vector r can be calculated using the cross product of s and
s_0 as

$$r = \frac{s \times s_0}{\|s\|^2} \tag{2.3}$$

It is noticed that both Eqs. (2.2) and (2.3) imply $\|s\| \neq 0$. When $\|s\| = 0$, the screw is defined to have *infinite pitch*, which has the form

$$S = \begin{bmatrix} 0 \\ hs \end{bmatrix} \tag{2.4}$$

where the dual part hs represents the direction vector instead.

2.2.1 Screw Interchange Operation

The general screw definition given in Eq. (2.1) is actually described using the Plücker ray coordinate frame [12]. This form of a screw is mostly used to represent a wrench in mechanism analysis. There also exists another form of a screw which exchanges the first and second part of the screw in Eq. (2.1). This new form of a screw is said to be represented in a Plücker axis coordinate frame, which can be written as

$$S_{axis} = \begin{bmatrix} s_0 \\ s \end{bmatrix} \tag{2.5}$$

In contrast to the Plücker ray coordinated screw, mostly S_{axis} is used to represent a twist in the mechanism analysis. The relationship between S_{axis} and S can be written as

$$S_{axis} = \Delta S \tag{2.6}$$

where Δ is the elliptical polar operator [13] which has the form

$$\Delta = \begin{bmatrix} 0 & I_3 \\ I_3 & 0 \end{bmatrix} \tag{2.7}$$

Δ has some properties as the following

$$\Delta = \Delta^{-1}$$
$$\Delta = \Delta^{T} \tag{2.8}$$
$$\Delta\Delta = I_3$$

2.2.2 Screw Coordinate Transformation

Since a screw can be represented in the form of a six-dimensional vector, it follows certain rules of coordinate transformation when its based coordinate frame

Fig. 2.2 Coordinate
transformation of a screw

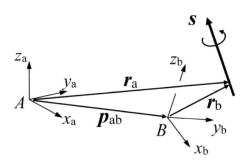

changes. Two Cartesian coordinate frames $\{A, x_a, y_a, z_a\}$ and $\{B, x_b, y_b, z_b\}$ are used to demonstrate the coordinate transformation of a screw, as shown in Fig. 2.2. Assume the symbol of a screw is $\mathbf{S}_a$ in the coordinate frame $\{A, x_a, y_a, z_a\}$, then it has the form as

$$\mathbf{S}_a = \begin{bmatrix} \mathbf{s}_a \\ \mathbf{s}_{a0} \end{bmatrix} = \begin{bmatrix} \mathbf{s}_a \\ \mathbf{r}_a \times \mathbf{s}_a + h\mathbf{s}_a \end{bmatrix} \tag{2.9}$$

Similarly, assume $\mathbf{S}_b$ is the symbol of $\mathbf{S}$ in the coordinate frame $\{B, x_b, y_b, z_b\}$ and it can be written as

$$\mathbf{S}_b = \begin{bmatrix} \mathbf{s}_b \\ \mathbf{s}_{b0} \end{bmatrix} = \begin{bmatrix} \mathbf{s}_b \\ \mathbf{r}_b \times \mathbf{s}_b + h\mathbf{s}_b \end{bmatrix} \tag{2.10}$$

where both $\mathbf{S}_a$ and $\mathbf{S}_b$ are written using Plücker ray coordinates. To develop the relationship between $\mathbf{S}_a$ and $\mathbf{S}_b$, we begin by looking into $\mathbf{s}_a$ and $\mathbf{s}_b$. According to the definition of a screw, both $\mathbf{s}_a$ and $\mathbf{s}_b$ are 3×1 direction vectors, thus one can be obtained by just rotating the other one according to the rotation matrix between two coordinate frames, which can be yielded as

$$\mathbf{s}_a = \mathbf{R}_{ab}\mathbf{s}_b \tag{2.11}$$

where $\mathbf{R}_{ab}$ is the 3×3 rotation matrix from $\{A, x_a, y_a, z_a\}$ to $\{B, x_b, y_b, z_b\}$. In order to develop the relationship between the second parts $\mathbf{S}_{a0}$ and $\mathbf{S}_{b0}$, firstly we utilize the position vector relationship as

$$\mathbf{r}_a = \mathbf{p}_{ab} + \mathbf{R}_{ab}\mathbf{r}_b \tag{2.12}$$

where $\mathbf{p}_{ab}$ is the translation vector from $\{A, x_a, y_a, z_a\}$ to $\{B, x_b, y_b, z_b\}$. Then the relationship between $\mathbf{S}_{a0}$ and $\mathbf{S}_{b0}$ can be derived as

$$\begin{aligned} \mathbf{r}_a \times \mathbf{s}_a + h\mathbf{s}_a &= (\mathbf{p}_{ab} + \mathbf{R}_{ab}\mathbf{r}_b) \times \mathbf{s}_a + h\mathbf{s}_a \\ &= \mathbf{p}_{ab} \times \mathbf{R}_{ab}\mathbf{s}_b + \mathbf{R}_{ab}\mathbf{r}_b \times \mathbf{R}_{ab}\mathbf{s}_b + h\mathbf{R}_{ab}\mathbf{s}_b \end{aligned} \tag{2.13}$$

Since it holds for the following formula as

$$\mathbf{R}_{ab}\mathbf{r}_b \times \mathbf{R}_{ab}\mathbf{s}_b = \mathbf{R}_{ab}(\mathbf{r}_b \times \mathbf{s}_b) \tag{2.14}$$

then Eq. (2.13) can be simplified as

$$
\begin{aligned}
\mathbf{r}_a \times \mathbf{s}_a + h\mathbf{s}_a &= \mathbf{p}_{ab} \times \mathbf{R}_{ab}\mathbf{s}_b + \mathbf{R}_{ab}(\mathbf{r}_b \times \mathbf{s}_b) + h\mathbf{R}_{ab}\mathbf{s}_b \\
&= \mathbf{p}_{ab} \times \mathbf{R}_{ab}\mathbf{s}_b + \mathbf{R}_{ab}(\mathbf{r}_b \times \mathbf{s}_b + h\mathbf{s}_b) \\
&= \mathbf{p}_{ab} \times \mathbf{R}_{ab}\mathbf{s}_b + \mathbf{R}_{ab}\mathbf{s}_{b0}
\end{aligned}
\tag{2.15}
$$

and this further leads to

$$\mathbf{s}_{a0} = \mathbf{p}_{ab} \times \mathbf{R}_{ab}\mathbf{s}_b + \mathbf{R}_{ab}\mathbf{s}_{b0} \tag{2.16}$$

According to Eqs. (2.11) and (2.16), we can obtain the relationship between $\mathbf{S}_a$ and $\mathbf{S}_b$ as

$$\mathbf{S}_a = \mathbf{Ad}_{ab}\mathbf{S}_b \tag{2.17}$$

where $\mathbf{Ad}_{ab}$ is the adjoint transformation matrix [14] and it has the form

$$\mathbf{Ad}_{ab} = \begin{bmatrix} \mathbf{R}_{ab} & \mathbf{0} \\ \mathbf{P}_{ab}\mathbf{R}_{ab} & \mathbf{R}_{ab} \end{bmatrix} \tag{2.18}$$

where $\mathbf{R}_{ab}$ is the 3×3 rotation matrix from $\{A, x_a, y_a, z_a\}$ to $\{B, x_b, y_b, z_b\}$, $\mathbf{P}_{ab}$ is the anti-symmetric matrix of translation vector $\mathbf{p}_{ab}$, it can be written as the following

$$\mathbf{P}_{ab} = \begin{bmatrix} 0 & -p_z & p_y \\ p_z & 0 & -p_x \\ -p_y & p_x & 0 \end{bmatrix} \tag{2.19}$$

where $\mathbf{p}_{ab} = \begin{bmatrix} p_x & p_y & p_z \end{bmatrix}^{\mathrm{T}}$. Further $\mathbf{Ad}_{ab}$ has the following properties as

$$
\begin{aligned}
\mathbf{Ad}_{ab} &= \Delta(\mathbf{Ad}_{ab}^{-\mathrm{T}})\Delta \\
\mathbf{Ad}_{ab}^{\mathrm{T}} &= \Delta(\mathbf{Ad}_{ab}^{-1})\Delta
\end{aligned}
\tag{2.20}
$$

As a result, Eq. (2.17) gives us the general form of screw coordinate transformation using the adjoint matrix $\mathbf{Ad}_{ab}$, and its properties are further given in Eq. (2.20).

2.2.3 Reciprocal Product of Two Screws

Following the formulation of screw coordinate transformation, the next is to discuss the relationships between two screws in the same coordinate frame. Among which the most important one is the *reciprocal product* [4]. It is formally defined by von Mises et al. [15] and has the form as

$$S_1 \circ S_2 = S_1^{\mathrm{T}} \Delta S_2 = s_1^{\mathrm{T}} s_{20} + s_2^{\mathrm{T}} s_{10} \tag{2.21}$$

From Eq. (2.21) we can see the reciprocal product of two screws is the sum of two multiplications between the first part of one screw to the second part of the other screw. It can be proved the reciprocal product between two screws is coordinate invariant. Assume a coordinate transformation $\mathbf{Ad}_{ab}$ is applied to the two screws S_1 and S_2, then we have

$$
\begin{aligned}
(\mathbf{Ad}_{ab}S_1) \circ (\mathbf{Ad}_{ab}S_2) &= (\mathbf{Ad}_{ab}S_1)^{\mathrm{T}} \Delta (\mathbf{Ad}_{ab}S_2) \\
&= S_1^{\mathrm{T}} (\mathbf{Ad}_{ab}^{\mathrm{T}} \Delta \mathbf{Ad}_{ab}) S_2
\end{aligned}
\tag{2.22}
$$

According to the properties of $\mathbf{Ad}_{ab}$ provided in Eq. (2.19), we have

$$
\begin{aligned}
\mathbf{Ad}_{ab}^{\mathrm{T}} \Delta \mathbf{Ad}_{ab} &= (\Delta \mathbf{Ad}_{ab}^{-1} \Delta) \Delta \mathbf{Ad}_{ab} \\
&= \Delta (\mathbf{Ad}_{ab}^{-1}) \mathbf{Ad}_{ab} \\
&= \Delta
\end{aligned}
\tag{2.23}
$$

Thus Eq. (2.22) can be further simplified as

$$(\mathbf{Ad}_{ab}S_1) \circ (\mathbf{Ad}_{ab}S_2) = S_1^{\mathrm{T}} \Delta S_2 = S_1 \circ S_2 \tag{2.24}$$

which suggests the reciprocal product of two screws is coordinate invariant. Actually this can also be explained by the physical interpretation of the instantaneous power done by a wrench with respect to a twist, which is also coordinate invariant. Normally we say a screw S_1 is reciprocal to the other screw S_2 if their reciprocal product is zero. The reciprocal relationship between two or more screws has been frequently used in the analysis of mechanisms. For example, it can be applied to identifying the constraints of a robotic system for a given degree of freedom and vice versa. Apart from the reciprocal relationship, a pair of screws whose reciprocal product is not equal to zero has also been investigated in the extension study of screw theory [6]. In that paper, the concept of repelling screws are used to define a pair of screws that have a positive reciprocal product, while the concept of contrary screws is used to define a screw pair that has negative reciprocal product. Together with the definition of reciprocal screws, they conclude the most general relationships between two screws.

2.3 Definition of the Screw System

A *screw system* is an aggregate of all screws which is symbolized as $\mathbb{S}$. Consider a set of n screws that are described in the same coordinate frame, then $\mathbb{S}$ has the form as

$$\mathbb{S} = \{ \boldsymbol{S}_1, \boldsymbol{S}_2, \dots, \boldsymbol{S}_n \} \tag{2.25}$$

This set of screws is said to be linearly independent if none of the screws can be written as a combination of the rest $n - 1$ screws. For example, the r-th screw $\boldsymbol{S}_r$ cannot be written as a linear combination of the remaining $(n - 1)$ screws as

$$\boldsymbol{S}_r \neq \lambda_1 \boldsymbol{S}_1 + \cdots + \lambda_{r-1} \boldsymbol{S}_{r-1} + \lambda_{r+1} \boldsymbol{S}_{r+1} + \cdots + \lambda_n \boldsymbol{S}_n \tag{2.26}$$

Further this n independent screws can expand a space, in which an arbitrary screw can be written as a combination of these n screws as

$$\boldsymbol{S} = \lambda_1 \boldsymbol{S}_1 + \lambda_2 \boldsymbol{S}_2 + \cdots + \lambda_n \boldsymbol{S}_n \tag{2.27}$$

which can also be written in a matrix form as

$$\boldsymbol{S} = \begin{bmatrix} \boldsymbol{S}_1 & \boldsymbol{S}_2 & \cdots \boldsymbol{S}_n \end{bmatrix} \begin{bmatrix} \lambda_1 \\ \lambda_2 \\ \vdots \\ \lambda_n \end{bmatrix} \tag{2.28}$$

Since each screw $\boldsymbol{S}_i (i = 1, \dots, n)$ is a 6×1 vector, $n \leq 6$ can guarantee the matrix $\begin{bmatrix} \boldsymbol{S}_1 & \boldsymbol{S}_2 & \cdots \boldsymbol{S}_n \end{bmatrix}$ has full rank. It is noticed that the reciprocal product of two screws is discussed in Sect. 2.2.3, followed by the discussion of reciprocal screws and repelling screws. As a correspondence, the related reciprocal screw system and repelling screw system are discussed subsequently.

2.3.1 Reciprocal Screw System

The reciprocal screw system is introduced first. According to the definition of reciprocal screws, a reciprocal screw system of a screw or screw system is composed of screws that are reciprocal to that screw or the screws in the given screw system. If $\mathbb{S}^r$ is utilized to represent the reciprocal screw system, then the following definition is given as

$$\mathbb{S}^r = \{ \boldsymbol{S}_1^r, \dots, \boldsymbol{S}_{6-n}^r \mid (\boldsymbol{S}_i^r \circ \boldsymbol{S}_j = 0, i = 1, \dots, 6 - n), \forall \boldsymbol{S}_j, j = 1, \dots, n \} \tag{2.29}$$

where the dimension of the given screw system is symbolized as $\dim(\mathbb{S})$ and that of the reciprocal screw system is $\dim(\mathbb{S}^r)$, they have the following relationship

$$\dim(\mathbb{S}) + \dim(\mathbb{S}^r) = 6 \tag{2.30}$$

The concept of a screw system and the reciprocal screw system is widely used in describing the twist or wrench space for a given mechanism or robotic platform. For example, for a serial robot, its twist space can be treated as a screw system whose components represent the motions of joints; in contrast, the reciprocal screw system illustrates the constraints of this serial robotic system that limits its motions.

2.3.2 Repelling Screw System

Similar to the reciprocal screw system, a new type of screw system can be established according to the extended concept of repelling screws. For a given screw system $\mathbb{S}$, its repelling screw system is symbolized as $\mathbb{S}^p$ where the superscript p means repelling. Any repelling screw $\boldsymbol{S}_k^p$ that belongs to $\mathbb{S}^p$ is defined to have a positive reciprocal product with $\boldsymbol{S}_j$ when $k = j$ and its reciprocal product with other screws are zero. This can be described as

$$\boldsymbol{S}_k^p = \begin{cases} \boldsymbol{S}_k^p \circ \boldsymbol{S}_j = 1, k = j \\ \boldsymbol{S}_k^p \circ \boldsymbol{S}_j = 0, k \neq j \end{cases} \tag{2.31}$$

As a result, we can always find n repelling screws for a given screw system to establish the repelling screw system

$$\mathbb{S}^p = \left\{ \boldsymbol{S}_1^p, \boldsymbol{S}_2^p, \ldots, \boldsymbol{S}_n^p \right\} \tag{2.32}$$

also the dimension of repelling screw system is equal to that of the given screw system

$$\dim(\mathbb{S}^p) = \dim(\mathbb{S}) = n \tag{2.33}$$

Further it will be shown repelling screws are particularly useful in the force analysis of a compliant mechanism in Chap. 8.

2.4 Definition of the Twist and Wrench

The definition of a screw is given in Sect. 2.2, which is given as a pure line vector without any physical meaning. Actually the exploration of the physical meanings of a screw can be traced to the work of Chasles and Poinsot in the early 1800s.

Chasles proved that a rigid motion in three dimensions can be represented by a rotation about an axis followed by a translation parallel to that line. This motion is generally referred as a *screw motion*. The infinitesimal version of a screw motion is called a *twist*, it can be used to describe the instantaneous velocity or the infinitesimal displacement of a rigid body in terms of its linear and angular components.

Poinsot discovered that any system of forces acting on a rigid body can be replaced by a single force applied along an axis, combined with a torque about the same axis. Such a force is referred as a *wrench*. Wrenches are dual to twists so that many of the theorems which apply to twists can be extended to wrenches. As a result, screw theory can be directly used to study the kinematics and statics for a given mechanism or robotic system.

2.4.1 Definition of a Twist and the Twist Space

In a three dimensional Euclidean space, the instantaneous velocity of a rigid body can be described using a twist, which can be written as

$$t = \begin{bmatrix} \boldsymbol{\omega} \\ \boldsymbol{v} \end{bmatrix} \tag{2.34}$$

where t is a 6×1 vector whose primary part $\boldsymbol{\omega} = \begin{bmatrix} \omega_x & \omega_y & \omega_z \end{bmatrix}^{\mathrm{T}}$ is a 3×1 angular velocity vector and the second part $\boldsymbol{v} = \begin{bmatrix} v_x & v_y & v_z \end{bmatrix}^{\mathrm{T}}$ is a 3×1 linear velocity vector. Since a twist can be described using a screw, it can be written as

$$t = \begin{bmatrix} \boldsymbol{\omega} \\ r \times \boldsymbol{\omega} + h\boldsymbol{\omega} \end{bmatrix} = \|\boldsymbol{\omega}\| \begin{bmatrix} s \\ r \times s + hs \end{bmatrix} = \|\boldsymbol{\omega}\| S \tag{2.35}$$

Thus a twist t can also be identified as a screw S associated with magnitude $\|\boldsymbol{\omega}\|$. In Eq. (2.35), r is the position vector of twist t, which can be calculated according to the calculation law of a screw as

$$r = \frac{\boldsymbol{\omega} \times \boldsymbol{v}}{\|\boldsymbol{\omega}\|^2} \tag{2.36}$$

Similarly h is the pitch of twist t, it has the form as

$$h = \frac{\boldsymbol{\omega}^{\mathrm{T}} \boldsymbol{v}}{\|\boldsymbol{\omega}\|^2} \tag{2.37}$$

When the rigid body is subject to a pure translation velocity $\boldsymbol{v}$, then it is said the twist t has an infinite pitch and its direction vector s is along the direction of $\boldsymbol{v}$, which has the form as

$$t = \begin{bmatrix} \mathbf{0} \\ \boldsymbol{\upsilon} \end{bmatrix} = \|\boldsymbol{v}\| \begin{bmatrix} \mathbf{0} \\ \boldsymbol{s} \end{bmatrix} \tag{2.38}$$

The concept of a twist representing the instantaneous velocity of a rigid body can be extended to represent its infinitesimal displacement for an infinitesimal period of time. For a time interval $[0, \Delta t]$, assume Δt is small enough, thus the orientation and position of twist t remain the same and the infinitesimal displacement of a rigid body with the twist motion can be written as

$$\int_0^{\Delta t} t \, dt = \int_0^{\Delta t} \|\boldsymbol{\omega}\| \boldsymbol{S} dt = \|\boldsymbol{\omega} \Delta t\| \boldsymbol{S} \tag{2.39}$$

Without the lose of generality, we use the same symbol t to represent the infinitesimal displacement vector as

$$t = \begin{bmatrix} \boldsymbol{\theta} \\ \boldsymbol{\delta} \end{bmatrix} = \|\boldsymbol{\theta}\| \begin{bmatrix} \boldsymbol{s} \\ \boldsymbol{s_0} \end{bmatrix} = \|\boldsymbol{\theta}\| \begin{bmatrix} \boldsymbol{s} \\ \boldsymbol{r} \times \boldsymbol{s} + h\boldsymbol{s} \end{bmatrix} \tag{2.40}$$

where $\|\boldsymbol{\omega} \Delta t\| = \|\boldsymbol{\theta}\|$, the primary part $\boldsymbol{\theta} = \begin{bmatrix} \theta_x & \theta_y & \theta_z \end{bmatrix}^{\mathrm{T}}$ is a 3×1 rotational displacement vector and the second part $\boldsymbol{\delta} = \begin{bmatrix} \delta_x & \delta_y & \delta_z \end{bmatrix}^{\mathrm{T}}$ is a 3×1 translational displacement vector. Following the definition of a twist, the concept of screw system can be extended to represent a group of twists, which is named as the *twist system*. A *twist space* is an aggregate of all defined twists, which can be utilized to represent the allowed motion of a rigid body, thus it is also called *freedom space*. If the twist system is symbolized using $\mathbb{S}$, then $\mathbb{S}$ has the form as

$$\mathbb{S} = \{t_1, t_2, \ldots, t_n\} \tag{2.41}$$

where a set of twists $t_i (i = 1, \ldots, n)$ are included in the twist system and they are described using the Plücker ray coordinates. If they are assumed to be linearly independent, then any allowed motion of the rigid body t is a linear combination of them and can be written as

$$t = t_1 + t_2 + \cdots + t_n \tag{2.42}$$

Since each twist t_i can be represented as an associated screw $\boldsymbol{S}_i$ with the magnitude θ_i according to Eq. (2.40), thus Eq. (2.42) can also be written in a matrix form as

$$t = \begin{bmatrix} \boldsymbol{S}_1 & \boldsymbol{S}_2 & \cdots & \boldsymbol{S}_n \end{bmatrix} \begin{bmatrix} \theta_1 \\ \theta_2 \\ \vdots \\ \theta_n \end{bmatrix} \tag{2.43}$$

It is noticed t is presented in the Plücker ray coordinates by far, sometimes it is more convenient to represent the infinitesimal displacement of a rigid body using the Plücker axis coordinates, which can be obtained by interchanging the first and second part of T as

$$T = \Delta t = \begin{bmatrix} \delta \\ \theta \end{bmatrix} \tag{2.44}$$

As a result, the allowed instantaneous motion or displacement of a rigid body in Plücker axis coordinates can be written as a sum of the n linearly independent twists according to Eqs. (2.43) and (2.44) as

$$T = \Delta \begin{bmatrix} S_1 & S_2 & \cdots & S_n \end{bmatrix} \begin{bmatrix} \theta_1 \\ \theta_2 \\ \vdots \\ \theta_n \end{bmatrix} \tag{2.45}$$

2.4.2 Definition of a Wrench and the Wrench Space

Similarly, a spatial force applied at the robotic platform can be represented using a wrench in screw theory. It contains a linear component (pure force) and an angular component (pure moment), which has the form as

$$w = \begin{bmatrix} f \\ m \end{bmatrix} \tag{2.46}$$

where w is a 6×1 vector whose primary part $f = \begin{bmatrix} f_x & f_y & f_z \end{bmatrix}^\mathrm{T}$ is a 3×1 force vector and the second part $m = \begin{bmatrix} m_x & m_y & m_z \end{bmatrix}^\mathrm{T}$ is a 3×1 moment vector. According to Poinsot's theorem, w can be described as a force applied along its axis and a torque about the same axis, which can be written as

$$w = \begin{bmatrix} f \\ r \times f + hf \end{bmatrix} = \|f\| \begin{bmatrix} s \\ r \times s + hs \end{bmatrix} = \|f\|S \tag{2.47}$$

Thus a wrench w can be identified as a screw S associated with intensity $\|w\|$. In Eq. (2.47), r is the position vector of wrench w and can be calculated as

$$r = \frac{f \times m}{\|f\|^2} \tag{2.48}$$

Similarly h is the pitch of wrench $\boldsymbol{w}$, it has the form as

$$h = \frac{\boldsymbol{f}^{\mathrm{T}}\boldsymbol{m}}{\|\boldsymbol{f}\|^2} \tag{2.49}$$

When the rigid body is subject to a pure torque $\boldsymbol{m}$, it is said the wrench $\boldsymbol{w}$ has infinite pitch and its direction vector $\boldsymbol{s}$ is along the direction of $\boldsymbol{m}$, which has the form as

$$\boldsymbol{w} = \begin{bmatrix} \boldsymbol{0} \\ \boldsymbol{m} \end{bmatrix} = \|\boldsymbol{m}\| \begin{bmatrix} \boldsymbol{0} \\ \boldsymbol{s} \end{bmatrix} \tag{2.50}$$

Following the definition of a wrench, the concept of a screw system can be extended to represent a group of wrenches, which can be named as *wrench system*. A *wrench space* is an aggregate of all defined wrench, which can be utilized to represent the forces that apply at a rigid body. If the wrench system is symbolized as $\mathbb{S}^r$, then $\mathbb{S}^r$ has the form as

$$\mathbb{S}^r = \{\boldsymbol{w}_1, \boldsymbol{w}_2, \ldots, \boldsymbol{w}_m\} \tag{2.51}$$

where a set of twists $\boldsymbol{w}_j (j = 1, \ldots, m)$ are included in the wrench system and they are described using the Plücker ray coordinates. If they are further linearly independent, then any applied load on a rigid body $\boldsymbol{w}$ is a linear combination of them and can be written as

$$\boldsymbol{w} = \boldsymbol{w}_1 + \boldsymbol{w}_2 + \cdots + \boldsymbol{w}_m \tag{2.52}$$

Since each twist $\boldsymbol{w}_j$ can be represented as an associated screw $\boldsymbol{S}_j$ with the magnitude f_j according to Eq. (2.47), thus Eq. (2.52) can also be written in a matrix form as

$$\boldsymbol{w} = \begin{bmatrix} \boldsymbol{S}_1 & \boldsymbol{S}_2 & \cdots & \boldsymbol{S}_m \end{bmatrix} \begin{bmatrix} f_1 \\ f_2 \\ \vdots \\ f_m \end{bmatrix} \tag{2.53}$$

It is noticed $\boldsymbol{w}$ is presented in the Plücker ray coordinates by far. In addition, a wrench can also be written using Plücker axis coordinates as

$$\boldsymbol{W} = \Delta\boldsymbol{w} = \begin{bmatrix} \boldsymbol{m} \\ \boldsymbol{f} \end{bmatrix} \tag{2.54}$$

As a result, the form of $\boldsymbol{W}$ can be further written according to Eqs. (2.53) and (2.54) as

$$\boldsymbol{W} = \Delta \begin{bmatrix} \boldsymbol{S}_1 & \boldsymbol{S}_2 & \cdots & \boldsymbol{S}_m \end{bmatrix} \begin{bmatrix} f_1 \\ f_2 \\ \vdots \\ f_m \end{bmatrix} \tag{2.55}$$

2.4.3 Example: The Twist and Wrench Space of a Compliant Parallel Mechanism

In this section, one example is given to illustrate the twist and wrench space of a compliant parallel mechanism. Some concepts about the screw representation of flexible elements and compliant mechanisms are given in advance merely for a purpose of demonstration, and readers can refer the following Chaps. 3 and 4 for the detailed explanations.

The discussed compliant mechanism is presented in Fig. 2.3, which has a base around and a functional platform in the center. This functional platform is elastically supported by 5 slender-beam type flexures, each of which is assumed to only resist translation along its axis but no other forms of motion. According to the constraint-based design principle, one flexible element can be represented using a constraint wrench along its axis direction. For example, $\boldsymbol{w}_1$ is utilized to represent the force exerted by the 1-th flexible element, it has the form

$$\boldsymbol{w}_1 = \begin{bmatrix} 1 & 0 & 0 & 0 & 0 & 0 \end{bmatrix}^{\mathrm{T}} \tag{2.56}$$

Fig. 2.3 One example of the freedom space and wrench space of a compliant parallel mechanism

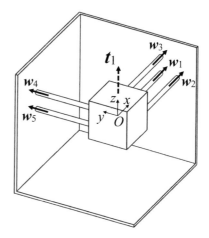

Further for all 5 flexible elements, their resistive forces form a constraint wrench space $\mathbb{S}^r$, where the superscript r means *reciprocal*. $\mathbb{S}^r$ can be written as

$$
\mathbb{S}^r = \begin{cases}
\boldsymbol{w}_1 = \begin{bmatrix} 1 & 0 & 0 & 0 & 0 & 0 \end{bmatrix}^{\mathrm{T}} \\
\boldsymbol{w}_2 = \begin{bmatrix} 1 & 0 & 0 & 0 & 0 & 1 \end{bmatrix}^{\mathrm{T}} \\
\boldsymbol{w}_3 = \begin{bmatrix} 1 & 0 & 0 & 0 & 1 & 0 \end{bmatrix}^{\mathrm{T}} \\
\boldsymbol{w}_4 = \begin{bmatrix} 0 & 1 & 0 & 0 & 0 & 0 \end{bmatrix}^{\mathrm{T}} \\
\boldsymbol{w}_5 = \begin{bmatrix} 0 & 1 & 0 & 1 & 0 & 0 \end{bmatrix}^{\mathrm{T}}
\end{cases}
\tag{2.57}
$$

Based on the wrench space $\mathbb{S}^r$, the allowed motion of the compliant parallel mechanism can be determined according to the reciprocal relationship discussed in Sect. 2.3.1

$$
\mathbb{S} \circ \mathbb{S}^r = \boldsymbol{0}
\tag{2.58}
$$

where $\mathbb{S}$ symbolizes the freedom space of the compliant mechanism. According to Eqs. (2.57) and (2.58), we can easily calculate the allowed motion of the compliant mechanism as

$$
\mathbb{S} = \left\{ \boldsymbol{t}_1 = \begin{bmatrix} 0 & 0 & 0 & 0 & 0 & 1 \end{bmatrix}^{\mathrm{T}} \right.
\tag{2.59}
$$

which is a translation along the z-axis of the global coordinate frame $\{O, x, y, z\}$ shown in Fig. 2.3. To verify this conceptual layout design of the compliant parallel mechanism, a finite element simulation is conducted to check its mobility under external loads. As shown in Fig. 2.4, the layout of the compliant parallel platform follows the one shown in Fig. 2.3. A translation along the z-axis of the global

Fig. 2.4 Finite element simulation of the deformation of the compliant parallel mechanism

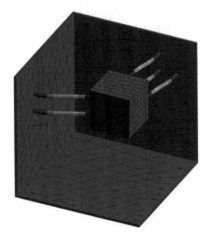

coordinate frame is added to the functional platform. From the result we can see the functional platform can move vertically following the allowed motion according to the constraint arrangement, thus validates the reciprocal relationship between the freedom space and wrench space defined early in this section.

2.5 Conclusions

In conclusion, this chapter gives a general introduction of screw theory and the important definitions and calculations within the framework. As two important concepts, twist and wrench are introduced subsequently in forms of both screw and screw system, and they are utilized to represent the freedom and constraint space of a compliant parallel mechanism as a preliminary example. This representation will be further utilized to analyze both flexible elements and their integrations in the following chapters.

References

1. Ball, R.S.: A Treatise on the Theory of Screws. Cambridge University Press (1900)
2. Dimentberg, F.M.: The screw calculus and its applications in mechanics. Technical Report, DTIC Document (1968)
3. Hunt, K.H.: Kinematic Geometry of Mechanisms. Clarendon Press, Oxford (1990)
4. Dai, J.S., Jones, J.R.: Null–space construction using cofactors from a screw–algebra context. Proc. Royal Soc. Lond. Ser. A: Math. Phys. Eng. Sci. **458**(2024), 1845–1866 (2002)
5. Dai, J.S., Huang, Z., Lipkin, H.: Mobility of overconstrained parallel mechanisms. J. Mech. Des. **128**, 220 (2006)
6. Ohwovoriole, M., Roth, B.: An extension of screw theory. J. Mech. Des. **103**(4), 725–735 (1981)
7. Blanding, D.L.: Exact constraint: machine design using kinematic processing. American Society of Mechanical Engineers (1999)
8. Hopkins, J.B.: Design of flexure-based motion stages for mechatronic systems via freedom, actuation and constraint topologies (FACT). Ph.D. thesis, Massachusetts Institute of Technology (2010)
9. Su, H.-J., Tari, H.: Realizing orthogonal motions with wire flexures connected in parallel. J. Mech. Des. **132**, 121002 (2010)
10. Yu, J.J., Li, S.Z., Qiu, C.: An analytical approach for synthesizing line actuation spaces of parallel flexure mechanisms. J. Mech. Des. **135**(12), 124501–124501 (2013)
11. Dai, J.S.: Geometrical Foundations and Screw Algebra for Mechanisms and Robotics. Higher Education Press, Beijing (2014). ISBN 9787040334838. (translated from Dai, J.S.: Screw Algebra and Kinematic Approaches for Mechanisms and Robotics. Springer, London (2016))
12. Plucker, J.: On a new geometry of space. Philos. Trans. R. Soc. Lond. **155**, 725–791 (1865)
13. Dai, J.S.: Finite displacement screw operators with embedded Chasles motion. J. Mech. Robot. **4**(4), 041002 (2012)
14. Dai, J.S.: Euler-rodrigues formula variations, quaternion conjugation and intrinsic connections. Mech. Mach. Theory **92**, 144–152 (2015)
15. Von Mises, R., Baker, E., Wohlhart, K.: Motor Calculus, a New Theoretical Device for Mechanics. Institute for Mechanics, University of Technology Graz (1996)

Chapter 3
Screw Representation of Flexible Elements

3.1 Introduction

Flexible elements are essential components which provide the deformation and flexibility of compliant mechanisms. Initially flexible elements are widely used in traditional rigid-body mechanisms and robots, most of which are in forms of simple springs such as the extensional and torsional springs. Generally, they are installed in either a translational or a revolute joint to resist motion in one direction but no other forms of motion, thus they are identified as single degree-of-freedom (DOF) flexible elements. For single DOF flexible elements such as simple springs, their spatial stiffness characteristics have relatively simple forms, which can be represented using a screw [1] and corresponding stiffness/compliance coefficients that can be found in standard mechanical textbooks [2].

In designing compliant mechanisms, researchers develop flexible elements with similar functions inspired by the application of simple springs. Successful applications include the beam-type flexures [3, 4] and blades [5]. However, due to the fact that compliant mechanisms integrate flexible elements without any rigid-joint connections, these flexures normally demonstrate flexibilities in more than one degree-of-freedom. Thus they are identified as multiple DOF flexible elements. Various approaches have been proposed to describe their compliance performance in different directions, such as the Euler-Bernoulli beam theory or classical beam theory, and the virtual joint model [6, 7]. Since it is revealed the virtual joint model also adopts the stiffness matrix of the cantilever beam from structural mechanics, the Euler-Bernoulli beam theory is used to model the compliance performance of multi-DOF flexible elements in this chapter.

C. Qiu and J. S. Dai, *Analysis and Synthesis of Compliant Parallel Mechanisms—Screw Theory Approach*, Springer Tracts in Advanced Robotics 139, https://doi.org/10.1007/978-3-030-48313-5_3

3.2 Single DOF Flexible Elements

For a flexible element, the infinitesimal displacement and applied force are linked to each other through its internal compliance/stiffness. For example, when an external wrench $\boldsymbol{w}$ is applied at a functional platform, it generates a corresponding displacement $\boldsymbol{T}$ due to the compliance of the flexible element, which can be written as

$$\boldsymbol{T} = \mathbf{C}\boldsymbol{w} \tag{3.1}$$

where $\mathbf{C}$ is the compliance matrix of the flexible element. If wrench $\boldsymbol{w}$ is written in the Plücker ray coordinates and twist $\boldsymbol{T}$ is written in Plücker axis coordinates, then $\mathbf{C}$ is positive or semi-positive definite for the unloaded configuration or the deflection of the flexible element is small [8]. Similarly, when an external displacement load $\boldsymbol{T}$ is applied, the flexible element generates a resulted reaction force $\boldsymbol{w}$ due to its stiffness, it can be written as

$$\boldsymbol{w} = \mathbf{K}\boldsymbol{T} \tag{3.2}$$

Both $\mathbf{C}$ and $\mathbf{K}$ reveal the intrinsic elastic performance of flexible elements, which are used on different occasions. For example, a compliance matrix $\mathbf{C}$ is generally used in analyzing a flexible element in a serial configuration while the stiffness matrix $\mathbf{K}$ is normally used in a parallel configuration.

Following the mechanism-equivalence principle, the flexible elements used in traditional mechanisms are introduced first, most of which are single degree-of-freedom (DOF) flexible elements such as the linear spring and torsional spring. For a linear spring, a linear infinitesimal deformation δ from its balanced position results in a resistive force f along the axis, which can be written as

$$f = k_\delta \delta \tag{3.3}$$

where k_δ is the stiffness coefficient of the linear spring and its inverse $c_\delta = k_\delta^{-1}$ is the compliance coefficient. For a torsional spring, there exists a similar formula. When a rotational infinitesimal displacement θ is applied from its balanced position, a resistive moment m about the axis of the torsional spring will be generated as

$$m = k_\theta \theta \tag{3.4}$$

where k_θ is the stiffness coefficient of the torsional spring and its inverse $c_\theta = k_\theta^{-1}$ is the corresponding compliance coefficient. Normally a linear spring is assembled in a translational joint and a torsional spring is assembled in a revolute joint.

3.2.1 Compliance Matrices of Single DOF Flexible Elements

Figure 3.1 shows a revolute joint installed with a torsional spring. A global coordinate frame $\{O, x, y, z\}$ is defined which has an arbitrary position and orientation. In $\{O, x, y, z\}$, the rotational axis of the revolute joint has a direction vector $\boldsymbol{n}$, whose position can be represented by a position vector $\boldsymbol{r}$. When an external load $\boldsymbol{w}$ is applied at one end of the revolute joint, it only generates a rotation θ about the axis $\boldsymbol{n}$ through the rotational compliance c_m but no other motions, which can be written as

$$\|\boldsymbol{\theta}\| = c_m \left[(\boldsymbol{r} \times \boldsymbol{n})^{\mathrm{T}} \quad \boldsymbol{n}^{\mathrm{T}} \right] \begin{bmatrix} \boldsymbol{f} \\ \boldsymbol{m} \end{bmatrix} \tag{3.5}$$

Since the displacement twist $\boldsymbol{T}$ is a rotation about axis $\boldsymbol{n}$, it has the form as

$$\boldsymbol{T} = \begin{bmatrix} \boldsymbol{\delta} \\ \boldsymbol{\theta} \end{bmatrix} = \|\boldsymbol{\theta}\| \begin{bmatrix} \boldsymbol{r} \times \boldsymbol{n} \\ \boldsymbol{n} \end{bmatrix} \tag{3.6}$$

Substituting Eq. (3.5) into Eq. (3.6), we can develop the relationship between $\boldsymbol{T}$ and $\boldsymbol{w}$ as

$$\boldsymbol{T} = c_\theta \begin{bmatrix} \boldsymbol{r} \times \boldsymbol{n} \\ \boldsymbol{n} \end{bmatrix} \left[(\boldsymbol{r} \times \boldsymbol{n})^{\mathrm{T}} \quad \boldsymbol{n}^{\mathrm{T}} \right] \boldsymbol{w} \tag{3.7}$$

from which we can further obtain the compliance matrix of the revolute joint as

$$\mathbf{C}_\theta = c_\theta \begin{bmatrix} \boldsymbol{r} \times \boldsymbol{n} \\ \boldsymbol{n} \end{bmatrix} \left[(\boldsymbol{r} \times \boldsymbol{n})^{\mathrm{T}} \quad \boldsymbol{n}^{\mathrm{T}} \right] \tag{3.8}$$

Assume the screw associated with the revolute joint is $\boldsymbol{S}$, then according to the definition of a screw, $\boldsymbol{S} = \left[\boldsymbol{n}^{\mathrm{T}} \quad (\boldsymbol{r} \times \boldsymbol{n})^{\mathrm{T}} \right]^{\mathrm{T}}$, and the compliance matrix $\mathbf{C}_\theta$ has a more general form as

$$\mathbf{C}_\theta = c_\theta (\Delta \boldsymbol{S})(\Delta \boldsymbol{S})^{\mathrm{T}} \tag{3.9}$$

Fig. 3.1 Torsional spring in a serial configuration

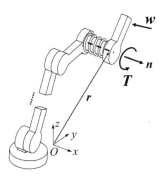

Fig. 3.2 Linear spring in a
serial configuration

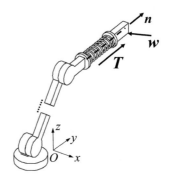

The compliance matrix of a prismatic joint associated with a linear spring can be
developed in a similar manner. As shown in Fig. 3.2, in the global coordinate frame
$\{O, x, y, z\}$, the translational axis of the prismatic joint has a direction vector n, when
an external load w is applied at the connector link, it will only generate a stroke δ
along the axis n but no other motions through the linear compliance c_δ, which can
be written as

$$\|\delta\| = c_\delta \begin{bmatrix} n^\mathrm{T} & 0^\mathrm{T} \end{bmatrix} \begin{bmatrix} f \\ m \end{bmatrix} \tag{3.10}$$

The displacement twist T is a translation along the axis n, it has the form as

$$T = \begin{bmatrix} \delta \\ 0 \end{bmatrix} = \|\delta\| \begin{bmatrix} n \\ 0 \end{bmatrix} = c_\delta \begin{bmatrix} n \\ 0 \end{bmatrix} \begin{bmatrix} n^\mathrm{T} & 0^\mathrm{T} \end{bmatrix} w \tag{3.11}$$

Accordingly, the compliance matrix of the prismatic joint in serial configuration
can be written as

$$\mathbf{C}_\delta = c_\delta \begin{bmatrix} n \\ 0 \end{bmatrix} \begin{bmatrix} n^\mathrm{T} & 0^\mathrm{T} \end{bmatrix} \tag{3.12}$$

which can be further simplified as

$$\mathbf{C}_\delta = c_\delta (\Delta S)(\Delta S)^\mathrm{T} \tag{3.13}$$

where S is the associated screw of the translational joint and $S = \begin{bmatrix} 0^\mathrm{T} & n^\mathrm{T} \end{bmatrix}^\mathrm{T}$.

3.2.2   Stiffness Matrices of Single DOF Flexible Elements

Conversely, when in a parallel configuration, stiffness matrices of single degree-of-
freedom flexible elements are developed rather than their compliance matrices. Let's

consider a prismatic joint assembled with a linear spring first. Assume an external displacement load w is applied at the connector link, then the prismatic joint will only generate a force f along the axis n but no other forces through the linear stiffness k_δ, which can be written as

$$\|f\| = k_\delta \left[n^{\mathrm{T}} \quad (r \times n)^{\mathrm{T}} \right] \begin{bmatrix} \delta \\ \theta \end{bmatrix} \tag{3.14}$$

Then the relationship between the resulted wrench w and external twist T can be written as

$$w = \begin{bmatrix} f \\ m \end{bmatrix} = \|f\| \begin{bmatrix} n \\ r \times n \end{bmatrix} = k_\delta \begin{bmatrix} n \\ r \times n \end{bmatrix} \left[n^{\mathrm{T}} \quad (r \times n)^{\mathrm{T}} \right] T \tag{3.15}$$

Thus the stiffness matrix $\mathbf{K}_\delta$ of the prismatic joint in a parallel configuration can be written as

$$\mathbf{K}_\delta = k_\delta \begin{bmatrix} n \\ r \times n \end{bmatrix} \left[n^{\mathrm{T}} \quad (r \times n)^{\mathrm{T}} \right] \tag{3.16}$$

which can be further simplified as

$$\mathbf{K}_\delta = k_\delta S S^{\mathrm{T}} \tag{3.17}$$

where the screw associated with the revolute joint is S and $S = \left[n^{\mathrm{T}} \quad (r \times n)^{\mathrm{T}} \right]^{\mathrm{T}}$ (Fig. 3.3).

Similarly, for a revolute joint, assuming an external displacement load T is applied at the connector link, it will only generate a moment m about the axis n but no other forces through the torsional stiffness k_θ, then the stiffness matrix of the translational joint has the form as

$$\mathbf{K}_\theta = k_\theta S S^{\mathrm{T}} \tag{3.18}$$

Fig. 3.3 Linear spring in parallel configuration

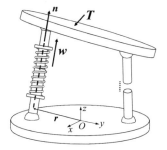

Table 3.1 Comparisons of compliance/stiffness of simple springs

Spring type	Serial configuration	
	Compliance matrix	Screw
Linear	$c_\delta(\Delta S)(\Delta S)^T$	$[\mathbf{0}^T \quad \mathbf{n}^T]^T$
Torsional	$c_\theta(\Delta S)(\Delta S)^T$	$[\mathbf{n}^T \quad (\mathbf{r} \times \mathbf{n})^T]^T$
Spring type	Parallel configuration	
	Stiffness matrix	Screw
Linear	$k_\delta \mathbf{S}\mathbf{S}^T$	$[\mathbf{n}^T \quad (\mathbf{r} \times \mathbf{n})^T]^T$
Torsional	$k_\theta \mathbf{S}\mathbf{S}^T$	$[\mathbf{0}^T \quad \mathbf{n}^T]^T$

where $\mathbf{S}$ is the associated screw of the translational joint and $\mathbf{S} = \begin{bmatrix} \mathbf{0}^T & \mathbf{n}^T \end{bmatrix}^T$. Further Table 3.1 gives a complete list of the compliance/stiffness matrices of simple springs in both serial and parallel configurations.

3.3 Multiple DOF Flexible Elements

Apart from the single DOF flexible elements such as linear springs and torsional springs, recently more types of flexible elements have been developed and utilized to design compliant mechanisms, such as the slender beams and blades. Unlike simple DOF flexible elements, these types of flexible elements usually generate coupled deflection other than single DOF motion based on their structural stiffness/compliance, thus they are also called multiple DOF flexible elements. As such, it is necessary to develop their corresponding compliance/stiffness matrices to determine the relationship between applied loads and resulted deflections.

Figure 3.4 presents a blade-type flexible element and its deflection with one end fixed to the base. The blade has a length of L, a width of b and a thickness of h. A coordinate frame $\{O, x, y, z\}$ is located as the center of the blade in its unloaded configuration. This coordinate frame is fixed in the space with its axes coincident with the principal axes of the blade. There is another body coordinate frame $\{O', x', y', z'\}$ which is fixed to the body of blade. It coincides with the coordinate frame $\{O, x, y, z\}$ when the blade is at its initial position and moves to a new position when the blade is deformed.

Under an external load, the blade deforms and generates a displacement in the space, including both the rotation and translation component. Under a small-deflection assumption, coordinate frames $\{O, x, y, z\}$ and $\{O', x', y', z'\}$ are assumed to coincide with each other, thus both the external load and spatial displacement can be described using the wrench and twist discussed in Sect. 2.4, and they both can be described in the global coordinate frame $\{O, x, y, z\}$. The relationship between the external load and spatial displacement can be expressed similarly using the stiffness or compliance matrix of the blade as simple flexible elements, which are symmetric

Fig. 3.4 A blade-type
compliance element

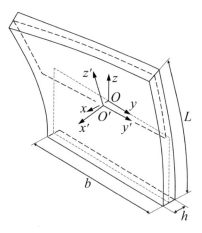

according to the small-deflection assumption [9]. Though complicated formulations have been proposed to address the force/deflection relationships in different directions, as have been discussed in [9], here for a purpose of demonstration, we mainly use the classic beam-deformation formulae and develop their corresponding stiffness/compliance matrices in the framework of screw theory.

As mentioned earlier, in contrast to single degree-of-freedom flexible elements, these beam-type flexible elements are treated as joints with multiple degrees of freedom, the number of which is determined by their stiffness/compliance matrices. For example, as shown in Fig. 3.4, the blade has a length of L, a width of b and a thickness of h. Assume $L \gg h$ and $b \gg h$, then it is reasonable to guess the blade is easy to rotate about y-axis and z-axis than x-axis; also it is easy to generate linear displacement along the x-axis other than y-axis and z-axis with the same magnitude of external load. If we treat the translation along x-axis and rotations about y-axis and z-axis as degrees of freedom and other motions as constraints, then the blade in Fig. 3.4 is equal to a compliant joint with three degrees of freedom. This is generally true for most blade-type flexible elements. Apart from blades, slender beams are another type of flexible elements that are widely used. It will be shown a slender beam has more degrees of freedom compared to a blade, which will also be discussed in the following sections.

3.3.1 Compliance/Stiffness Matrices of Slender Beams

The slender-beam type flexible elements are investigated first in this section. A slender beam model is presented in Fig. 3.5, whose bottom end is fixed to the ground and the other end is connected to a cubic-shape functional platform for the external load. The length of the slender beam is L and its diameter is D, where $L \gg D$ for the slender beam assumption. As indicated in [10, 11], stiffness matrix of a beam

Fig. 3.5 A beam-type
flexible element

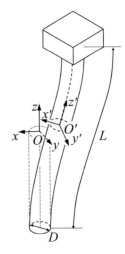

element can be reduced to a 6×6 matrix by introducing zero displacement boundary conditions at one end so the displacements on the other end are relative deflections. Assume the beam is un-deformed with a constant cross section, and a coordinate frame is attached at the free end of the beam, a stiffness matrix K can be developed to depict the relationship between an external load and the small deflection. This K has non-zero coupling terms. If we further define a coordinate frame with the origin coincident with the center of the beam, either by using screw coordinate transformation [11] or the transport theorem in structural mechanics [10], we can obtain a stiffness matrix that is diagonal without coupling terms. Here a coordinate frame $\{O, x, y, z\}$ is located as the center of the beam in its unloaded position. This coordinate frame is fixed in the space with its axes coincident with the principal axes of the beam. There is another body coordinate frame $\{O', x', y', z'\}$ which is fixed to the beam body. It coincides with the coordinate frame $\{O, x, y, z\}$ when the beam is at its initial position and moves to a new position when it is deformed.

When a small deformation occurs at the end of the slender beam under an external load, both the deformation and the load can be described using a twist T and a wrench w. Generally, twist T contains both linear and rotational displacement components and it is written in the Plücker axis coordinates. The wrench w contains both force and moment components and it is written in the Plücker axis coordinates. The compliance matrix C can be derived according to Eq. (3.1) once we know the relationship between T and w in all six directions (three rotations and three translations). Since the coordinate frame is coinciding with the mid-span of the beam and only small deformation is considered, the compliance matrix can be derived according to the classic Euler-Bernoulli beam theory, which has a simplified diagonal form as

$$\mathbf{C} = diag \begin{bmatrix} c_1 & c_2 & c_3 & c_4 & c_5 & c_6 \end{bmatrix}$$
$$= diag \begin{bmatrix} \frac{L^3}{12EI_y} & \frac{L^3}{12EI_x} & \frac{L}{EA} & \frac{L}{EI_x} & \frac{L}{EI_y} & \frac{L}{GI_z} \end{bmatrix} \tag{3.19}$$

where A is the cross-sectional area of the beam and $A = \pi r^2$. The shear effect and inertia of rotation of the beam section are ignored. Moment of inertias I_x and I_y are given as $I_x = I_y = \frac{1}{4}\pi r^4$, and the torsion constant $I_z = \frac{1}{2}\pi r^4$. For slender beams with other cross-sections, their moment of inertia and torsion constant can be found in [12] and here they are omitted. E and G are Young's modulus and shear modulus, they are varied according to the selection of materials.

In contrast to the single DOF flexible elements which have inconsistent compliance/stiffness matrices, for a slender beam, its stiffness matrix is simply the inverse of its compliance matrix, which can be written as

$$\mathbf{K} = \mathbf{C}^{-1} = diag \begin{bmatrix} \frac{12EI_y}{L^3} & \frac{12EI_x}{L^3} & \frac{EA}{L} & \frac{EI_x}{L} & \frac{EI_y}{L} & \frac{GI_z}{L} \end{bmatrix} \tag{3.20}$$

which is due to the fact that the slender beam is assumed to have the same boundary condition in either serial configuration or parallel configuration, as will be discussed in the integration of flexible elements in the next chapter.

3.3.1.1   Degrees of Freedom of Slender Beams

Further, the developed compliance/stiffness matrix can help us decide the number of degree-of-freedom of a slender-beam flexible element, which depends on the comparison of magnitudes of deflection in different directions. To achieve this requires a unification of the units of compliance elements in the compliance matrix. It is noticed that the unit of the compliance matrix $\mathbf{C}$ obtained in Eq. (3.19) has different units of the first three diagonal elements compared to those of the second three diagonal elements. Thus in order to make the comparison meaningful, a preliminary procedure is conducted by unifying the units of components in both external load $\boldsymbol{w}$ and deformation twist $\boldsymbol{T}$. In accordance with this, a characteristics matrix $\mathbf{L}$ is introduced which has a form as

$$\mathbf{L} = diag \begin{bmatrix} x & x & x & 1 & 1 & 1 \end{bmatrix} \tag{3.21}$$

where x is the characteristics length, generally it has the same order of magnitude of the length of beam L. Then $\mathbf{L}$ is utilized to unify the units of $\boldsymbol{w}$ and $\boldsymbol{T}$, which can be written as

$$\boldsymbol{w}' = \mathbf{L}\boldsymbol{w}$$
$$\boldsymbol{T}' = \mathbf{L}^{-1}\boldsymbol{T} \tag{3.22}$$

According to the definition of $\boldsymbol{w}$ and $\boldsymbol{T}$, it is easy to identify that $\boldsymbol{w}'$ has a unified unit (N m) while $\boldsymbol{T}'$ has a unified unit (rad). Substituting the modified $\boldsymbol{w}'$ and $\boldsymbol{T}'$ into Eq. (3.1), we can obtain

$$\boldsymbol{L}\boldsymbol{T}' = \boldsymbol{C}\boldsymbol{L}^{-1}\boldsymbol{w}' \tag{3.23}$$

from which we can obtain the modified compliance matrix $\boldsymbol{C}'$ as

$$\boldsymbol{C}' = \boldsymbol{L}^{-1}\boldsymbol{C}\boldsymbol{L}^{-1} \tag{3.24}$$

Further substituting the formulations of $\boldsymbol{C}$ and $\boldsymbol{L}$ into Eq. (3.24), the parameterized $\boldsymbol{C}'$ can be developed as

$$\boldsymbol{C}' = diag \left[\frac{L^3}{12EI_y} \cdot \frac{1}{x^2} \; \frac{L^3}{12EI_x} \cdot \frac{1}{x^2} \; \frac{L}{EA} \cdot \frac{1}{x^2} \; \frac{L}{EI_x} \; \frac{L}{EI_y} \; \frac{L}{GI_z} \right] \tag{3.25}$$

which can be further simplified according to the formulations of I_x, I_y and I_z

$$\boldsymbol{C}' = \left(\frac{L}{12EI_y} \right) \cdot diag \left[\left(\frac{L}{x} \right)^2 \; \left(\frac{L}{x} \right)^2 \; 3\left(\frac{r}{L} \right)^2 \cdot \left(\frac{L}{x} \right)^2 \; 12 \; 12 \; 12(1+v) \right] \tag{3.26}$$

In terms of x, usually, it has the same order of magnitude of L. As such, it can be identified from Eq. (3.26) that all diagonal elements $\boldsymbol{C}'(i, i)$ except for $\boldsymbol{C}'(3, 3)$ have the same order of magnitude and they are much bigger than $\boldsymbol{C}'(3, 3)$ (The ratio between $\boldsymbol{C}'(3, 3)$ and other components $\boldsymbol{C}'(i, i)(i \neq 3)$ has the same order of magnitude with $\frac{r^2}{L^2}$ which is smaller than 0.01). As a result, it is reasonable to conclude a slender beam has five degrees of freedom and one constraint which is along the longitudinal z-axis of the beam. The five degrees of freedom include two traverse translations δ_x and δ_y, and three rotations θ_x, θ_y and θ_z about the corresponding axes. Selected degrees of freedom are further shown in Fig. 3.6.

Fig. 3.6 The degrees of freedom of a slender-beam flexible element with external loads applying at the tip and the other end fixed at the base

3.3.2 Compliance/Stiffness Matrices of Blades

Apart from slender beams, blades are other types of widely used flexible elements. The biggest difference them lies in the fact they have different geometrical shapes which result in different degrees of freedom. For a slender beam, it has been shown to have five degrees of freedom, including two traverse translations and three rotations. For a blade, its degrees of freedom can be estimated following the same DOF-analysis procedure of slender beams.

Figure 3.4 already presents a blade-type flexible element. The blade has a length of L, a width of b and a thickness of h. The L and b are designed to have similar size while $L, b \gg h$. A coordinate frame $\{O, x, y, z\}$ is located as the center of the blade when it is undeformed. This coordinate frame is fixed in the space with its axes coincident with the principal axes of the blade. There is another body coordinate frame $\{O', x', y', z'\}$ which is fixed to the body of blade. It coincides with the coordinate frame $\{O, x, y, z\}$ when the blade is at its initial position and moves to a new position when the blade is deformed. Since the coordinate frame is coinciding with the mid-span of the beam and only small deformation is considered, the compliance matrix $\mathbf{C}$ can be derived according to the the classic Euler-Bernoulli beam theory, which has the same form as the one given in Eq. (3.19). However, the blade has a different shape, as it has a constant rectangular cross-section area of bh and a length of L. The shear effect and inertia of rotation of the blade section are ignored. Moment of inertias I_x and I_y are given as $I_x = \frac{1}{12}hb^3$ and $I_y = \frac{1}{12}bh^3$, and torsion constant I_z for the rectangular section is given by [12] as $I_z = bh^3(\frac{1}{3} - 0.21\frac{h}{b}(1 - \frac{h^4}{12b^4}))$ when $\frac{h}{b} \leq 1$. As a result, we can develop a similar unified compliance matrix $\mathbf{C}$ following Eq. (3.25), which can be written as

$$\mathbf{C}' = \left(\frac{L}{12EI_x}\right) \cdot diag\left[\left(\frac{b}{h}\right)^2 \cdot \left(\frac{L}{x}\right)^2 \left(\frac{L}{x}\right)^2 \left(\frac{b}{x}\right)^2 \cdot \left(\frac{L}{x}\right)^2 \; 12 \; 12\left(\frac{b}{h}\right)^2 \; 6(1+v)\left(\frac{b}{h}\right)^2\right]$$
(3.27)

from which we can compare the relative values between each diagonal compliance elements according to the ratios of dimensions. Particularly, L, b and x have the same order of magnitude, thus $\frac{L}{x} \approx 1$ and $\frac{b}{L} \approx 1$. However, $b \gg h$ thus $\mathbf{C}'(i, i)(i = 1, 5, 6)$ are much bigger compared to $\mathbf{C}'(i, i)(i = 2, 3, 4)$, they correspond to the translation along x-axis and rotations about y-axis and z-axis. As a result, the blade-type compliance element is identified to have three degrees of freedom, and they are presented in Fig. 3.7. It is worth noticing that some of the diagonal elements of the compliance matrix have been simplified, for example, we ignore the shear effects for the traverse stiffness. However, the formation of the diagonal matrix of the compliance matrix still validates if the coordinate frame is selected to coincide with the principal axes of the blade [9], with or without simplification of diagonal elements.

Fig. 3.7 The degrees of
freedom of a blade
compliance element

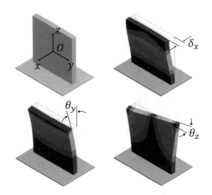

3.4   Coordinate Transformation of Flexible Elements

As has been shown above, the compliance/stiffness matrices of flexible elements
with both single and multiple degree-of-freedoms have been developed. Particularly, in the analysis of multi-DOF flexible elements, some special coordinate frames
are selected to develop a relatively simple formula for their compliance/stiffness
matrices. However, for a compliant mechanism which generally has more than one
flexible element, usually, a global coordinate frame is defined to describe the spatial
compliance performance of the whole platform. As a result, this requires the com-
pliance/stiffness matrix of one flexible element described in the localized coordinate
frame to be able to be described in another coordinate frame, which can be achieved
using the coordinate transformation.

Without the loss of generality, beam-type flexure is selected to demonstrate the
coordinate transformation of both its compliance matrix and stiffness matrix. As
shown in Fig. 3.8, the beam flexure has one end fixed to the ground and the other
end loaded by an external wrench w. When w is applied at the end of the beam, the
beam generates a deformation T. Two coordinate frames are presented in Fig. 3.8,
one is $\{A, x_a, y_a, z_a\}$ that coincides with the center of the beam, another one is
$\{B, x_b, y_b, z_b\}$ that coincides with the tip of the beam.

Fig. 3.8 Coordinate
transformation of the
compliance matrix of a
compliance element

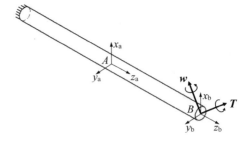

Assume a force/torque sensor is attached to the tip of the cantilever beam, then normally the external wrench w as well as the resulted deformation twist T are presented in the tip end coordinate frame $\{B, x_b, y_b, z_b\}$, which are symbolized as w_b and T_b, they have the following relationship as

$$T_b = C_b w_b \tag{3.28}$$

where C_b is the compliance matrix of the slender beam in the coordinate frame $\{B, x_b, y_b, z_b\}$. If we are going to study the performance of the cantilever beam in the center coordinate frame $\{A, x_a, y_a, z_a\}$, then both the external load and deformation twist are needed to be written in the new coordinate frame as w_a and T_a. The relationship between w_a and w_b can be written as

$$w_a = Ad_{ab} w_b \tag{3.29}$$

where both w_a and w_b are written using the Plücker ray coordinates and Ad_{ab} is the adjoint coordinate transformation matrix that has been given in Eq. (2.17), it is written here as

$$Ad_{ab} = \begin{bmatrix} R_{ab} & 0 \\ P_{ab}R_{ab} & R_{ab} \end{bmatrix} \tag{3.30}$$

where R_{ab} is the 3×3 rotation matrix from $\{A, x_a, y_a, z_a\}$ to $\{B, x_b, y_b, z_b\}$, P_{ab} is the anti-symmetric matrix representing the translation vector p_{ab} from $\{A, x_a, y_a, z_a\}$ to $\{B, x_b, y_b, z_b\}$. The deformation twist T due to the external force w is written using Plücker axis coordinates, for which we have a similar coordinate transformation formula as

$$(\Delta T_a) = Ad_{ab}(\Delta T_b) \tag{3.31}$$

According to the properties of Δ and Ad_{ab} shown in Eqs. (2.8) and (2.19), (3.31) can be further simplified as

$$T_a = (\Delta Ad_{ab} \Delta)T_b = Ad_{ab}^{-T} T_b \tag{3.32}$$

Substituting Eqs. (3.29) and (3.32) into Eq. (3.28), we can obtain

$$Ad_{ab}^{T} T_a = C_b Ad_{ab}^{-1} w_a \tag{3.33}$$

which can be further deduced as

$$T_a = Ad_{ab}^{-T} C_b Ad_{ab}^{-1} w_a \tag{3.34}$$

Since the compliance matrix C in the coordinate frame $\{A, x_a, y_a, z_a\}$ has the form $T_a = C_a w_a$, we can then get the relationship between C_a and C_b as

$$\mathbf{C}_a = \mathbf{Ad}_{ab}^{-T} \mathbf{C}_b \mathbf{Ad}_{ab}^{-1} \tag{3.35}$$

The coordinate transformation of the stiffness matrix of the slender beam can be developed in a similar way, or even simply by just reversing Eq. (3.35) as

$$\mathbf{C}_a^{-1} = \mathbf{Ad}_{ab} \mathbf{C}_b^{-1} \mathbf{Ad}_{ab}^{T} \tag{3.36}$$

According to the relationship between stiffness and compliance matrix $\mathbf{C} = \mathbf{K}^{-1}$, Eq. (3.35) can be further written as

$$\mathbf{K}_a = \mathbf{Ad}_{ab} \mathbf{K}_b \mathbf{Ad}_{ab}^{T} \tag{3.37}$$

As a result, the coordinate transformation of both compliance and stiffness matrix have been developed and they are given in Eqs. (3.35) and (3.37). It should be mentioned the coordinate transformation formulations developed here are also suitable for single DOF flexible elements, which lay the foundation for the further spatial integration of them in designing compliant mechanisms.

3.4.1 Example: Compliance Matrix of a Slender Beam in the Tip-End Coordinate Frame

In order to verify the coordinate transformation using screw theory, in this section, a detailed algebraic calculation is provided to develop the compliance matrix $\mathbf{C}_b$ in the tip coordinate frame $\{B, x_b, y_b, z_b\}$ from the compliance matrix $\mathbf{C}_a$ in the center coordinate frame $\{A, x_a, y_a, z_a\}$ which has a simple diagonal form. According to Eq. (3.35), $\mathbf{C}_b$ can be written in the form of a partitioned matrix as

$$\begin{aligned}
\mathbf{C}_b &= \mathbf{Ad}_{ab}^{T} \mathbf{C}_a \mathbf{Ad}_{ab} \\
&= \begin{bmatrix} \mathbf{R}_{ab}^{T} & (\mathbf{P}_{ab}\mathbf{R}_{ab})^{T} \\ \mathbf{0} & \mathbf{R}_{ab}^{T} \end{bmatrix} \begin{bmatrix} \mathbf{C}_{11} & \mathbf{0} \\ \mathbf{0} & \mathbf{C}_{22} \end{bmatrix} \begin{bmatrix} \mathbf{R}_{ab} & \mathbf{0} \\ \mathbf{P}_{ab}\mathbf{R}_{ab} & \mathbf{R}_{ab} \end{bmatrix} \\
&= \begin{bmatrix} \mathbf{C}_{11} + (\mathbf{P}_{ab}\mathbf{R}_{ab})^{T}\mathbf{C}_{22}(\mathbf{P}_{ab}\mathbf{R}_{ab}) & \mathbf{P}_{ab}^{T}\mathbf{C}_{22} \\ \mathbf{C}_{22}\mathbf{P}_{ab} & \mathbf{C}_{22} \end{bmatrix}
\end{aligned} \tag{3.38}$$

Since $\mathbf{C}_a$ is the compliance matrix of the slender beam in the coordinate frame $\{A, x_a, y_a, z_a\}$, it has the form as the one given in Eq. (3.19) where $\mathbf{C}_{11} = diag[c_1, c_2, c_3]$ and $\mathbf{C}_{22} = diag[c_4, c_5, c_6]$. As for $\mathbf{Ad}_{ab}$, it is the adjoint transformation matrix from $\{A, x_a, y_a, z_a\}$ to $\{B, x_b, y_b, z_b\}$, thus $\mathbf{R}_{ab} = \mathbf{I}$, and the translation vector $\boldsymbol{p}_{ab} = \begin{bmatrix} 0 & 0 & \frac{L}{2} \end{bmatrix}^{T}$ that leads to the anti-symmetric matrix $\mathbf{P}_{ab}$ as

$$\mathbf{P}_{ab} = \begin{bmatrix} 0 & -\frac{L}{2} & 0 \\ \frac{L}{2} & 0 & 0 \\ 0 & 0 & 0 \end{bmatrix} \tag{3.39}$$

Substituting Eq. (3.39) and the formula of $\mathbf{C}_a$ into Eq. (3.38), $\mathbf{C}_b$ can be written as

$$\mathbf{C}_b = \begin{bmatrix} c_1 + c_5 \left(\frac{L}{2}\right)^2 & 0 & 0 & 0 & c_5 \left(\frac{L}{2}\right) & 0 \\ 0 & c_2 + c_4 \left(\frac{L}{2}\right)^2 & 0 & -c_4 \left(\frac{L}{2}\right) & 0 & 0 \\ 0 & 0 & c_3 & 0 & 0 & 0 \\ 0 & -c_4 \left(\frac{L}{2}\right) & 0 & c_4 & 0 & 0 \\ c_5 \left(\frac{L}{2}\right) & 0 & 0 & 0 & c_5 & 0 \\ 0 & 0 & 0 & 0 & 0 & c_6 \end{bmatrix} \tag{3.40}$$

We can use the derived formula to calculate the deformation of beam-tip under external load. For example, assume a vertical force $\boldsymbol{w}_b$ is applied at the tip of the slender beam as

$$\boldsymbol{w}_b = \begin{bmatrix} f\ 0\ 0\ 0\ 0\ 0 \end{bmatrix}^{\mathrm{T}} \tag{3.41}$$

Then the corresponding deformation twist $\boldsymbol{T}_b$ can be calculated according to Eq. (3.40) by substituting the vertical load $\boldsymbol{w}_b$, which can be written as

$$\begin{aligned} \boldsymbol{T}_b &= \left[c_1 \cdot f + c_5 \left(\frac{L}{2}\right)^2 \cdot f\ 0\ 0\ 0\ c_5 \left(\frac{L}{2}\right) \cdot f\ 0 \right]^{\mathrm{T}} \\ &= \left[\frac{fL^3}{3EI_y}\ 0\ 0\ 0\ \frac{fL^2}{2EI}\ 0 \right]^{\mathrm{T}} \end{aligned} \tag{3.42}$$

which suggests the vertical force f along x-axis causes a translational deformation $(fL^3/3EI_y)$ along the x-axis and a rotation $(fL^2/2EI)$ about the y-axis, which is exactly the same results as those obtained using general structural-mechanics approaches.

3.5   Conclusions

In conclusion, this chapter describes various types of flexible elements in the framework of screw theory. It will be shown that various types of flexible elements are used in designing and analyzing compliant mechanisms in the following chapters, such as the shape memory alloy actuator in Chaps. 5 and 6, the slender beam elements in Chap. 7 and the origami creases in Chap. 8. For a more complete set of flexible elements, The readers can further refer [13]. Based on the effective degrees-of-freedom, both single-DOF flexible elements as well as multi-DOF flexible elements are introduced and their compliance/stiffness matrices are investigated systematically. Particularly an example is given to demonstrate the coordinate transformation of the developed compliance matrix in the framework of screw theory, which will be used to construct compliant mechanisms in the next chapter.

References

1. Huang, S., Schimmels, J.M.: The bounds and realization of spatial stiffnesses achieved with simple springs connected in parallel. IEEE Trans. Robot. Autom. **14**(3), 466–475 (1998)
2. Shigley, J.E., Mischke, C.R., Budynas, R.G., Liu, X., Gao, Z.: Mechanical Engineering Design, vol. 89. McGraw-Hill, New York (1989)
3. Parise, J.J., Howell, L.L., Magleby, S.P.: Ortho-planar linear-motion springs. Mech. Mach. Theory **36**(11), 1281–1299 (2001)
4. Awtar, S., Slocum, A.H.: Constraint-based design of parallel kinematic XY flexure mechanisms. J. Mech. Des. **129**(8), 816–830 (2007)
5. Blanding, D.L.: Exact constraint: machine design using kinematic processing. American Society of Mechanical Engineers (1999)
6. Pashkevich, A., Chablat, D., Wenger, P.: Stiffness analysis of overconstrained parallel manipulators. Mech. Mach. Theory **44**(5), 966–982 (2009)
7. Pashkevich, A., Klimchik, A., Chablat, D.: Enhanced stiffness modeling of manipulators with passive joints. Mech. Mach. Theory **46**(5), 662–679 (2011)
8. Duffy, J.: Statics and Kinematics with Applications to Robotics. Cambridge University Press (1996)
9. Hale, L.C.: Principles and techniques for designing precision machines. Technical Report, Lawrence Livermore National Lab., CA, US (1999)
10. Tuma, J.J.: Handbook of Structural and Mechanical Matrices. McGraw-Hill Inc., New York, NY, USA (1987)
11. Ciblak, N., Lipkin, H.: Design and analysis of remote center of compliance structures. J. Robot. Syst. **20**(8), 415–427 (2003)
12. Young, W.C., Budynas, R.G.: Roark's Formulas for Stress and Strain, vol. 6. McGraw-Hill, New York (2002)
13. Howell, L.L., Magleby, S.P., Olsen, B.M., Wiley, J.: Handbook of Compliant Mechanisms. Wiley Online Library (2013)

Chapter 4
Construction of Compliant Mechanisms

4.1 Introduction

Both serial and parallel configuration of compliant mechanisms are discussed in this chapter following the mechanism-equivalent principle. For a compliant serial mechanism, it is natural to establish its forward kinematics and inverse force analysis [1]. Based on them, the compliance matrix of the serial compliant mechanism with single-DOF flexible elements can be developed directly. In terms of multi-DOF flexible elements such as slender beams and blades, they occur compliance in more than one degree-of-freedom. As such, the compliance development of serial compliant mechanisms composed of multi-DOF flexible elements needs a modified procedure, which involves the coordinate transformation of individual flexible element's compliance matrix and their integration.

The same applies to parallel compliant mechanisms, though their inverse kinematics and forward force analysis are developed instead of the forward kinematics and inverse force analysis. These pave way for the stiffness construction of parallel compliant mechanisms with single-DOF flexible elements directly. Similarly, for parallel compliant mechanisms with multi-DOF flexible elements, a modified procedure is adopted which can properly integrate the stiffness matrices of individual flexible elements.

4.2 Serial Configuration of Compliant Mechanisms

The serial-configuration compliant mechanisms are discussed in this section. According to the mechanism-equivalence design principle, the traditional serial-type mechanism is introduced first. Figure 4.1 presents a typical serial-type traditional mechanism. It consists of a base, an end-effector and intermediate links connected by

© The Editor(s) (if applicable) and The Author(s), under exclusive license to Springer Nature Switzerland AG 2021
C. Qiu and J. S. Dai, *Analysis and Synthesis of Compliant Parallel Mechanisms—Screw Theory Approach*, Springer Tracts in Advanced Robotics 139, https://doi.org/10.1007/978-3-030-48313-5_4

Fig. 4.1 A
serial-configuration
mechanism with single DOF
flexible elements

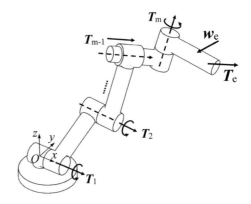

single-DOF joints. Assume there is a torsional stiffness in each revolute joint and
a linear stiffness in each prismatic joint, then each joint can be treated as a single-
DOF flexible element and the whole mechanical platform as a serial-type compliant
mechanism.

When an external load w_e is applied at the end-effector, each flexible element
generates a deformation $T_i (i = 1, \ldots, m)$ and together generates a displacement T_e
of the end-effector. The relationship between w_e and T_e is established through the
compliance matrix of the serial compliant mechanism. In order to obtain the compliance matrix, we first consider the relationship between the infinitesimal deformation
of the end effector T_e and that of compliance elements $T_i (i = 1, \ldots, m)$, which has
the form as

$$T_e = \sum_{i=1}^{m} T_i \tag{4.1}$$

which indicates T_e is the aggregation of T_i. The formula of T_i associated with
its corresponding screw S_i has been given in Eq. (3.6). Substituting Eq. (3.6) into
Eq. (4.1), we can obtain

$$T_e = \Delta J_s \delta\theta = \Delta \begin{bmatrix} S_1 & \cdots & S_m \end{bmatrix} \begin{bmatrix} \delta\theta_1 \\ \vdots \\ \delta\theta_m \end{bmatrix} \tag{4.2}$$

where J_S is the Jacobian matrix of the compliant mechanism, and both T_e and
$S_i (i = 1, \ldots, m)$ are 6×1 unit instantaneous deformation twists written in the
global coordinate frame $\{O, x, y, z\}$. The formulations of $S_i (i = 1, \ldots, m)$ are listed
in Table 3.1. $\delta\theta_i$ is the rotational angle of the i-th compliant joint if it is a revolute
joint or the linear displacement if it is a prismatic joint. They together integrates
the $m \times 1$ deformation vector $\delta\theta$. The deduction of Eq. (4.2) is also named as the
forward kinematics analysis. Following the forward kinematics analysis of the com-

pliant mechanism, the next step is to establish the relationship between external load w_e and resistive force of each flexible element according to the instant power relationship of the compliant mechanism, which has the form as

$$(w_e)^T T_e = \tau^T \delta\theta \tag{4.3}$$

where τ is a $m \times 1$ vector whose element τ_i is the magnitude of the resistive wrench w_i associated with the compliance element i. Substituting Eq. (4.2) into Eq. (4.3), the relationship between τ and w_e can be derived as

$$\tau = (\Delta J_S)^T w_e \tag{4.4}$$

which is identified as the inverse force analysis.

4.2.1 Compliant Serial Mechanisms with Single-DOF Flexible Elements

Based on the forward kinematics and inverse force analysis, the compliance matrix of this serial-type compliant mechanism with single-DOF flexible elements can be developed accordingly. For the i-th compliance joint, the relationship between the displacement and resistive force holds as

$$\delta\theta_i = c_{\theta,i}\tau_i \tag{4.5}$$

where $c_{\theta,i}$ is the corresponding compliance coefficient. Since Eq. (4.5) can be generalized to all m compliant joints, the relationship between $\delta\theta$ and τ can be developed in a matrix form as

$$\delta\theta = C_\theta \tau \tag{4.6}$$

where C_θ is a $m \times m$ symmetric matrix whose diagonal element is $c_{\theta,i} (i = 1, \ldots, m)$. For the whole compliant platform, we can establish the relationship between T_e and w_e as

$$T_e = C_e w_e \tag{4.7}$$

Substituting Eq. (4.2) into Eq. (4.7), the relationship between C_θ and C_e can be established initially as

$$C_e w_e = \Delta J_S C_\theta \tau \tag{4.8}$$

Further by substituting Eq. (4.4) into Eq. (4.8), we can obtain

$$\mathbf{C}_e = (\Delta \mathbf{J}_S)\mathbf{C}_\theta(\Delta \mathbf{J}_S)^{\mathrm{T}} \qquad (4.9)$$

which gives us the formula of the compliance matrix of a compliant mechanism. Further Eq. (4.4) can be expanded as

$$
\begin{aligned}
\mathbf{C}_e &= \begin{bmatrix} \Delta \mathbf{S}_1 & \cdots & \Delta \mathbf{S}_m \end{bmatrix}
\begin{bmatrix} c_{\theta,1} & & \\ & \ddots & \\ & & c_{\theta,m} \end{bmatrix}
\begin{bmatrix} (\Delta \mathbf{S}_1)^{\mathrm{T}} \\ \vdots \\ (\Delta \mathbf{S}_m)^{\mathrm{T}} \end{bmatrix} \\
&= \sum_{i=1}^{m} c_{\theta,i}(\Delta \mathbf{S}_i)(\Delta \mathbf{S}_i)^{\mathrm{T}}
\end{aligned}
\qquad (4.10)
$$

whose element is exactly the compliance matrix of a single flexible element listed the Table 3.1, which indicates the compliance matrix $\mathbf{C}_e$ of the serial compliant mechanism is the summation of flexible elements' compliance matrices.

4.2.2 Compliant Serial Mechanisms with Multi-DOF Flexible Elements

Apart from compliant mechanisms that are composed of single-DOF flexible elements, recently more types of compliant mechanisms are made using multi-DOF flexible elements such as slender beams and blades. It has been shown the formulations of the compliance matrices of multi-DOF flexible elements are different from those of single-DOF ones. As a result, the compliance development of this type of serial compliant mechanisms needs a modified procedure and it is presented in this section.

Figure 4.2 shows one compliant mechanism which is a serial chain of slender beams. It has one base fixed to the ground, a number of slender beams and intermediate stages that connect them, as well as an end-effector. A global coordinate frame $\{O, x, y, z\}$ is located at the base of the compliant mechanism. For the i-th slender beam, a local coordinate frame $\{O_i, x_i, y_i, z_i\}$ is located at the center of it,

Fig. 4.2 A serial-type compliant mechanism with multi-DOF compliance elements

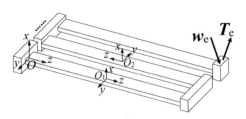

thus its compliance matrix $\mathbf{C}_i$ has the form as the one given in Eq. (3.19). For a purpose of simplicity, the intermediate stages are assumed rigid and their compliance are omitted. When an external load $\boldsymbol{w}_e$ is applied on the end-effector, it generates a deformation twist $\boldsymbol{T}_e$ which is the aggregation of each flexible element's deformation twist $\boldsymbol{T}_i (i = 1, \ldots, m)$. The relationship between $\boldsymbol{T}_e$ and $\boldsymbol{T}_i$ has been given in Eq. (4.1), where both $\boldsymbol{T}_e$ and $\boldsymbol{T}_i$ are described in the same global coordinate frame $\{O, x, y, z\}$. According to the coordinate transformation law, the deformation $\boldsymbol{T}_i$ of the i-th flexible element can be represented in the local coordinate frame $\{O_i, x_i, y_i, z_i\}$ and it is symbolized as $\boldsymbol{T}'_i$. The relationship between $\boldsymbol{T}_i$ and $\boldsymbol{T}'_i$ can be written as

$$\boldsymbol{T}'_i = \Delta \mathbf{Ad}_{ie} \Delta \boldsymbol{T}_i = \mathbf{Ad}_{ie}^{-\mathrm{T}} \boldsymbol{T}_i \tag{4.11}$$

where $\mathbf{Ad}_{ie} (i = 1, \ldots, m)$ is the adjoint transformation matrix between the local coordinate frame $\{O_i, x_i, y_i, z_i\}$ and the global coordinate frame $\{O, x, y, z\}$, it has the form

$$\mathbf{Ad}_{ie} = \begin{bmatrix} \mathbf{R}_{ie} & \mathbf{0} \\ \mathbf{P}_{ie}\mathbf{R}_{ie} & \mathbf{R}_{ie} \end{bmatrix} \tag{4.12}$$

Submitting Eq. (4.12) into Eq. (4.11), we can obtain

$$\boldsymbol{T}_e = \sum_{i=1}^{m} \mathbf{Ad}_{ie}^{\mathrm{T}} \boldsymbol{T}'_i \tag{4.13}$$

In contrast, the external load applied at the end effector is transmitting to each compliant element rather than balancing all the resistive forces of them, which has the form as

$$\boldsymbol{w}'_i = \mathbf{Ad}_{ie}\boldsymbol{w}_e \tag{4.14}$$

where $\boldsymbol{w}_e$ is the external load in the global coordinate frame, $\boldsymbol{w}'_i$ is the transmitted internal load applied at the i-th flexible element which is expressed in the local coordinate frame $\{O_i, x_i, y_i, z_i\}$. Both Eqs. (4.13) and (4.14) establish the relationship between quantities of end-effector and individual flexible elements; particularly, the relationship between the localized deformation and the internal load of each flexible element has been established through the compliance matrix, which is written as

$$\boldsymbol{T}'_i = \mathbf{C}_i \boldsymbol{w}'_i \tag{4.15}$$

where $\mathbf{C}_i$ is the compliance matrix of one slender beam and it has been given in Eq. (3.19). Similarly, for the whole compliant mechanism, we can also use a global compliance matrix $\mathbf{C}_e$ to establish the relationship between end-effector deformation and force as

$$\boldsymbol{T}_e = \mathbf{C}_e \boldsymbol{w}_e \tag{4.16}$$

Substituting Eqs. (4.15) and (4.16) into Eq. (4.13), we can obtain

$$\mathbf{C}_e \mathbf{w}_e = \sum_{i=1}^{m} \mathbf{Ad}_{ie}^{\mathrm{T}} \mathbf{C}_i \mathbf{w}_i' \tag{4.17}$$

which can be further deduced by substituting into Eq. (4.14), which results in

$$\mathbf{C}_e \mathbf{w}_e = \sum_{i=1}^{m} \mathbf{Ad}_{ie}^{\mathrm{T}} \mathbf{C}_i \mathbf{Ad}_{ie} \mathbf{w}_e \tag{4.18}$$

which suggests the relationship between $\mathbf{C}_e$ and $\mathbf{C}_i$ can be expressed as

$$\mathbf{C}_e = \sum_{i=1}^{m} \mathbf{Ad}_{ie}^{\mathrm{T}} \mathbf{C}_i \mathbf{Ad}_{ie} \tag{4.19}$$

Until now, the serial configuration of compliant mechanisms has been evaluated in terms of both single-DOF and multi-DOF flexible elements. Their corresponding compliance matrices are derived in Eqs. (4.10) and (4.19), from which we can see they are both linear combinations of compliance matrices of individual flexible elements.

4.3 Parallel Configuration of Compliant Mechanisms

Apart from the serial type of compliant mechanisms, there exists another type of compliant mechanisms which integrates flexible elements in a parallel configuration. This type of compliant mechanisms is identified as compliant parallel mechanisms. Unlike compliant serial mechanisms, compliant parallel mechanisms have properties such as high-payload, higher accuracy of motion, etc. They have been used in many industrial applications, such as picking and mounting robots in automation line in large scale as well as high accuracy manufacturing platform in micro-scale.

Different from the serial configuration of compliant mechanisms which emphasize the compliance performance, for compliant parallel mechanisms, their stiffness performance are of most importance. As such, we shall build up the stiffness matrix of a compliant parallel mechanism to address this issue. Figure 4.3 shows one compliant parallel mechanism that is composed of single-DOF flexible elements. Assuming an external displacement load $\mathbf{T}_e$ is applied at the platform, each supporting limb generates a corresponding displacement and results in a deformation of the flexible element in each limb. The process of calculating individual deformations of flexible elements from the displacement load $\mathbf{T}_e$ is identified as the inverse kinematics. Subsequently, each flexible element generates a resistive force and together generates a reaction force $\mathbf{w}_e$ of the whole platform. This process is identified as the forward force analysis. In order to develop the stiffness matrix of the compliant parallel mech-

Fig. 4.3 A compliant parallel mechanism with single-DOF flexible elements

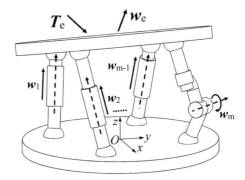

anism, we shall conduct both inverse kinematics analysis and forward force analysis first.

Following the analysis above, when an external displacement load is applied at the functional platform, each supporting limb generates a resistive wrench w_i and together integrates into the resistive wrench w_e of the whole platform, which can be written as

$$w_e = \sum_{i=1}^{m} w_i \tag{4.20}$$

Since each wrench w_i can be represented as a screw S_i associated with force magnitude f_i, Eq. (4.20) can be further written as

$$w_e = \mathbf{J}_S f = \begin{bmatrix} S_1 & \cdots & S_m \end{bmatrix} \begin{bmatrix} f_1 \\ \vdots \\ f_m \end{bmatrix} \tag{4.21}$$

where $\mathbf{J}_S$ is the Jacobian matrix of the compliant mechanism whose column is S_i ($i = 1, \ldots, m$). They are written in the global coordinate frame $\{O, x, y, z\}$. f is the $m \times 1$ vector whose element f_i is the force magnitude when the compliance element is a prismatic joint or torque magnitude for the rotational joint. According to the instant power relationship, we can obtain

$$(T_e)^{\mathrm{T}} w_e = \delta\theta^{\mathrm{T}} f \tag{4.22}$$

where $\delta\theta$ is a $m \times 1$ vector whose element $\delta\theta_i$ is the displacement magnitude of each compliance element. Substituting Eq. (4.21) into Eq. (4.22), the relationship between $\delta\theta$ and T_e can be derived as

$$\delta\theta = (\mathbf{J}_S)^{\mathrm{T}} T_e \tag{4.23}$$

this will be used in the following sections for stiffness construction of compliant parallel mechanisms.

4.3.1 Compliant Parallel Mechanisms with Single-DOF Flexible Elements

Similar to compliant serial mechanisms, the stiffness matrix of a compliant parallel mechanism that is composed of single degree-of-freedom flexible elements is developed first. For the i-th compliant joint, the relationship between the force magnitude and deformation magnitude can be described as

$$f_i = k_{\theta,i} \delta\theta_i \tag{4.24}$$

where $k_{\theta,i}$ is the corresponding stiffness coefficient. Equation (4.24) can be generalized to all m compliance joints. As a result, all m forces f_i and deformations $\delta\theta_i$ can be represented using $m \times 1$ vectors f and $\delta\theta$ and they have the following relationship

$$f = \mathbf{K}_\theta \delta\theta \tag{4.25}$$

where $\mathbf{K}_\theta$ is a $m \times m$ symmetric matrix whose diagonal element is $k_{\theta,i}$ ($i = 1, \ldots, m$). Similarly, we can establish the relationship between the external force w_e and the infinitesimal deformation of end-effector T_e as

$$w_e = \mathbf{K}_e T_e \tag{4.26}$$

Substituting Eqs. (4.25) and (4.26) into Eq. (4.21), the relationship between $\mathbf{K}_\theta$ and $\mathbf{K}_e$ can be established as

$$\mathbf{K}_e T_e = \mathbf{J}_S \mathbf{K}_\theta \delta\theta \tag{4.27}$$

Further by substituting Eq. (4.23) into Eq. (4.27), we can obtain

$$\mathbf{K}_e = (\mathbf{J}_S)\mathbf{K}_\theta(\mathbf{J}_S)^{\mathrm{T}} \tag{4.28}$$

Further Eq. (4.28) can be expanded as

$$\mathbf{K}_e = \begin{bmatrix} \mathbf{S}_1 & \cdots & \mathbf{S}_m \end{bmatrix} \begin{bmatrix} k_{\theta,1} & & \\ & \ddots & \\ & & k_{\theta,m} \end{bmatrix} \begin{bmatrix} \mathbf{S}_1^{\mathrm{T}} \\ \vdots \\ \mathbf{S}_m^{\mathrm{T}} \end{bmatrix} \qquad (4.29)$$

$$= \sum_{i=1}^{m} k_{\theta,i} \mathbf{S}_i \mathbf{S}_i^{\mathrm{T}}$$

which indicates $\mathbf{K}_e$ is the sum of $k_{\theta,i} \mathbf{S}_i \mathbf{S}_i^{\mathrm{T}}$ which is exactly the stiffness matrix of a single flexible element in a parallel configuration, it is also listed in the Table 3.1.

4.3.2 Compliant Parallel Mechanisms with Multi-DOF Flexible Elements

It can be seen Eq. (4.28) only suits when each flexible element has single degree-of-freedom. In terms of compliant parallel mechanisms that are supported with flexible elements such as slender beams or blades who have more than one degree-of-freedoms, a modified stiffness matrix development should be considered.

Figure 4.4 shows one compliant mechanism which is a parallel combination of multi-DOF flexible elements. Part of the flexible elements is shown in Fig. 4.4, including two slender beams on the sidewall and one blade on the bottom. They connect the centralized functional platform to the base around. For each flexible element, a local coordinate frame $\{O_i, x_i, y_i, z_i\}$ is located at its center with coordinate axes coinciding with its principal axes. A global coordinate frame $\{O, x, y, z\}$ is located at the center of the functional platform and it is fixed in the space.

When an external displacement load $\mathbf{T}_e$ is applied at the functional platform, each flexible element deforms following $\mathbf{T}_e$ and generates a resistive wrench $\mathbf{w}_i$, they together integrate into a resistive wrench $\mathbf{w}_e$ to balance $\mathbf{T}_e$. As a result, the force equilibrium presented in Eq. (4.20) for compliant parallel mechanism with single-DOF flexible elements still holds for those with multi-DOF compliance elements.

Fig. 4.4 A compliant parallel mechanism with multi DOF flexible elements

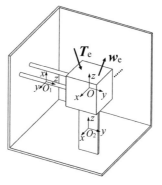

For the flexible element $i(i = 1, \ldots, m)$, the internal force $\boldsymbol{w}_i$ resulted from its deformation can also be described in the local coordinate frame $\{O_i, x_i, y_i, z_i\}$ as

$$\boldsymbol{w}_i^{'} = \mathbf{Ad}_{ie}\boldsymbol{w}_i \tag{4.30}$$

Where $\mathbf{Ad}_{ie}$ is the adjoint transformation matrix between local coordinate frame $\{O_i, x_i, y_i, z_i\}$ and global coordinate frame $\{O, x, y, z\}$, it has the same form as Eq. (4.12). Substituting Eq. (4.30) into Eq. (4.20), we can obtain

$$\boldsymbol{w}_e = \sum_{i=1}^{m} \mathbf{Ad}_{ie}^{-1}\boldsymbol{w}_i^{'} \tag{4.31}$$

Regarding the deformation $\boldsymbol{T}_e$ of the functional platform, when it deforms, each flexible element generates a corresponding deformation $\boldsymbol{T}_i$ with respect to the global coordinate frame. By introducing the coordinate transformation matrix $\mathbf{Ad}_{ie}$, we have

$$\boldsymbol{T}_i^{'} = \Delta\mathbf{Ad}_{ie}\Delta\boldsymbol{T}_e = \mathbf{Ad}_{ie}^{-T}\boldsymbol{T}_e \tag{4.32}$$

From Eqs. (4.31) and (4.32) we can see $\boldsymbol{w}_i^{'}$ and $\boldsymbol{T}_i^{'}$ are the generated force and deformation of the i-th flexible element described in the local coordinate frame $\{O_i, x_i, y_i, z_i\}$, they can be linked through the stiffness matrix as

$$\boldsymbol{w}_i^{'} = \mathbf{K}_i\boldsymbol{T}_i^{'} \tag{4.33}$$

where $\mathbf{K}_i$ is the stiffness matrix of the i-th flexible element and it has been given in Eq. (3.20). Similarly, we can define the global stiffness matrix $\mathbf{K}_e$ of the whole compliant parallel mechanism as

$$\boldsymbol{w}_e = \mathbf{K}_e\boldsymbol{T}_e \tag{4.34}$$

Substituting Eqs. (4.33) and (4.34) into Eq. (4.31), we have

$$\mathbf{K}_e\boldsymbol{T}_e = \sum_{i=1}^{m} \mathbf{Ad}_{ie}^{-1}\mathbf{K}_i\boldsymbol{T}_i^{'} \tag{4.35}$$

which can be further simplified by substituting into Eq. (4.32) and leads to the formulation of $\mathbf{K}_e$ as

$$\mathbf{K}_e = \sum_{i=1}^{m} \mathbf{Ad}_{ie}^{-1}\mathbf{K}_i\mathbf{Ad}_{ie}^{-T} \tag{4.36}$$

Up to now, compliant mechanisms with serial and parallel configurations that are constructed with both single-DOF and multi-DOF flexible elements have been

Table 4.1 Stiffness and compliance matrix comparisons of compliant mechanisms

	Serial configuration $\mathbf{C}_e$	Parallel configuration $\mathbf{K}_e$
Single DOF	$\sum_{i=1}^{m} c_{\theta,i}(\Delta \boldsymbol{S}_i)(\Delta \boldsymbol{S}_i)^{\mathrm{T}}$	$\sum_{i=1}^{m} k_{\theta,i} \boldsymbol{S}_i \boldsymbol{S}_i^{\mathrm{T}}$
Multi DOF	$\sum_{i=1}^{m} \mathbf{Ad}_{ie}^{\mathrm{T}} \mathbf{C}_i \mathbf{Ad}_{ie}$	$\sum_{i=1}^{m} \mathbf{Ad}_{ie}^{-1} \mathbf{K}_i \mathbf{Ad}_{ie}^{-\mathrm{T}}$

studied and their compliance/stiffness matrices have been developed accordingly in Sects. 4.2 and 4.3. Table 4.1 summarizes the corresponding formulations of compliant mechanisms in different configurations. It is worth noticing the provided formulae are based on some simplifications and assumptions. For example, in serial configurations, only compliances of flexible elements are considered; in parallel configurations, only stiffnesses of flexible elements are considered. When readers are referring the Table 4.1, it is suggested necessary modifications be adopted according to the real physical models.

Further Fig. 4.5 gives us a few examples of flexible elements discussed in Chap. 3 and their integrated compliant mechanisms, as have been discussed in this section. The flexible elements include a crease-type flexible element in Fig. 4.5a that has one rotational degree-of-freedom; its integration is an origami-type cartoon which is widely used in food packaging industry, as show in Fig. 4.5b. A spring-type flexible element is presented in Fig. 4.5c that has one translational degree-of-freedom, its integration is a compliant parallel platform that presents certain mobilities based on the configuration of integrated flexible limbs, as shown in Fig. 4.5d. Finally a beam-type flexible element is presented in Fig. 4.5e which demonstrates a spatial compliance, and its integration is a compliant planar spring shown in Fig. 4.5f which has been used in micro-value control. The selected examples clearly suggest that a compliant mechanism can be treated as an integration of flexible elements, and the mathematical framework proposed in Chap. 3 and this chapter can help us describe their compliance behaviours mathematically.

4.4 Problem Formulation of Compliant Mechanism Design

The presented study in Chap. 3 and this chapter pave the way for the further compliant mechanism designs in the framework of screw theory. Several typical design topics are covered in the following sections, ranging from the conceptual design to the dimensional design, from the analysis to the synthesis, as well as the large deflection problems.

Fig. 4.5 Multiple types of
flexible elements and their
integrated compliant
mechanisms

(a) A crease-type flexible element

(b) A crease-integrated compliant origami folds

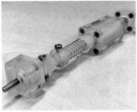

(c) A spring-type flexible element

(d) A compliant parallel platform

(e) A beam-type flexible element

(f) A compliant planar spring

4.4.1 Conceptual Design of Compliant Mechanisms

One of the key design problems is to arrange suitable flexible elements in a way
that can provide desired spatial compliance/stiffness performance. In the initial conceptual design stage, flexible elements are selected according to their degrees of
freedom rather than the detailed compliance/stiffness matrix, as can be seen from
the categorization of flexible elements in Chap. 3. This is identified as the conceptual design of compliant mechanisms. As has been shown in the example given in
Sect. 2.4.3, we can use twists to represent the preferred motions of a compliant mechanism and wrenches to represent the constraints exerted by the flexible elements. The
relationships between them can be written as

$$w^r \circ t = 0 \tag{4.37}$$

according to which the corresponding constraint space that is reciprocal to the freedom space can be determined, which can be further utilized to determine the layouts of actuators as well. A compliant parallel mechanism is developed accordingly using this conceptual design approach, which will be further discussed in Chap. 5.

4.4.2 Compliance/Stiffness Decomposition of Compliant Mechanisms

The initial concept design can help us obtain the layout of a compliant mechanism, from which we can construct its compliance/stiffness matrix and conduct further analysis and optimum design based on them. This is also identified the compliance/stiffness construction (analysis) problem. Apart from the analysis problem, another type of design problem exists, which is called the synthesis problem that aims to synthesize the pre-determined compliance/stiffness matrix and obtain the configuration of flexible elements. This is also identified as the mechanism decomposition process.

The importance of the decomposition problem lies within the fact that sometimes we don't necessarily know what configuration we need to construct a compliant mechanism but only know what compliance performance we prefer from it. As such, we may give an estimated compliance/stiffness matrix first and then decompose it according to their intrinsic properties. For instance, as listed in Table 4.1, it can be obtained that the stiffness matrix $\mathbf{K}_e$ has the following property as

$$tr(\mathbf{K}_e \Delta) = 0 \tag{4.38}$$

where $\mathbf{K}_e$ is the stiffness matrix and Δ is the elliptical polar operator [2] defined in Sect. 2.2.1. It will be shown in Chap. 6 that this property will be essential in conducting stiffness synthesis of compliant mechanisms. As a result, we can utilize these properties to decompose the stiffness matrix $\mathbf{K}_e$ and obtain the layout of flexible elements in return, and this is identified as the compliance/stiffness decomposition of compliant mechanisms.

4.4.3 Compliance/Stiffness Parameterization and Optimization of Compliant Mechanisms

Assuming we have determined the configuration of a compliant mechanism and its compliance/stiffness matrix, we can conduct further analysis such as parameterization and optimization. These advanced analyses enable us to evaluate the effect of every design parameters on the overall compliance/stiffness performance, which is particularly useful for the compliant-mechanism design with multi-DOF flexible elements such as slender beams and blades, whose compliance/stiffness matrices are fully determined by both geometrical and material properties. For example, a widely used compliant mechanism, the ortho-planar spring, will be utilized to demonstrate the parameterization and optimization design in Chap. 7. Its compliance matrix $\mathbf{C}_b$ is built based on the integration of slender beams, which turns out to be a function of design parameters $\mathbf{x}$, $\mathbf{y}$ as

$$\mathbf{C}_b = \mathscr{F}(\mathbf{x}, \mathbf{y}) \tag{4.39}$$

where $\mathbf{x}$ represents a vector of geometrical parameters such as b, h, etc, and $\mathbf{y}$ represents a vector of material parameters such as E, ν, etc, they can also be found in the compliance matrix formula of a slender beam in Eq. (3.19). Subsequently, we can utilize $\mathscr{F}(\mathbf{x}, \mathbf{y})$ to conduct further compliance analysis and optimization with respect to the design parameters $\mathbf{x}$, $\mathbf{y}$. This part of work is identified as compliance/stiffness parameterization and optimization of compliant mechanisms and the detailed study will be given in Chap. 7.

4.4.4 Large-Deformation Analysis of Compliant Mechanisms

The compliance/stiffness matrix based approach proves to be efficient and useful in designing compliant mechanisms in the unloaded configuration. However, the developed compliance/stiffness matrix may not be the same when a compliant mechanism undertakes relatively large external loads. Taking a compliant parallel mechanism for example, the force equilibrium formula has been given in Eq. 4.21 when it achieves force balance under an external load. Assume the external load $\mathbf{w}_e$ is increased by $\delta\mathbf{w}_e$, then $\delta\mathbf{w}_e$ can be obtained by taking the derivative of two sides of Eq. 4.21, which has the form as

$$\delta\mathbf{w}_e = \delta\mathbf{J}_s f + \mathbf{J}_s \delta f \tag{4.40}$$

Equation 4.40 indicates the stiffness matrix that describes the relationship between $\delta\mathbf{w}_e$ and δf is also determined by the internal force f. Only when f is relatively small or equal to zero then the stiffness matrix is what we have derived in previous sections, such as Eq. 4.28 and it is symmetric. Nevertheless, Eq. 4.21 still holds true even for a compliant mechanism under a large deflection. As such, it is more convenient to explore the large deformation of a compliant mechanism in forms of its force transmission properties. As have been discussed in Sects. 4.2 and 4.3, inverse force analysis has been utilized in compliant serial mechanisms while forward force analysis has been used in the compliant parallel mechanisms. For a general compliant mechanism that is a mixture of both configurations, there is a need to unify both forward and inverse force analysis in the same framework. For example, a compliant parallel mechanism can have supporting limbs which are serial combinations of flexible elements. In order to conduct force analysis of such types of compliant

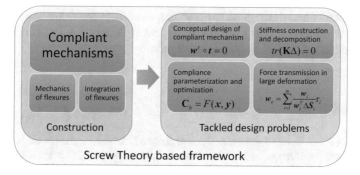

Fig. 4.6 Problem formulation of compliant mechanism design in the framework of screw theory

mechanisms, the concept of repelling-screw and repelling-screw system introduced in Sect. 2.3.2 are utilized, which leads to the formulation of force analysis of one compliant serial chain as

$$w_e = \sum_{i=1}^{m} \frac{w_i}{w_i^{\mathrm{T}} \Delta S_i} \tau_i \tag{4.41}$$

where S_i is the screw associated with i-th compliant joint and τ_i is the corresponding resistive force. w_e is the integrated resistive force of all compliant joints in the compliant serial mechanism. The detailed explanation of Eq. 4.41 will be given in Chap. 8. Then the force analysis of the whole compliant mechanism are developed in the same manner based on Eq. 4.41. As a result, we are able to address the large deformation of compliant mechanisms through the force analysis using the repelling-screw based approach.

Up to now, four different types of design problems of compliant mechanisms have been introduced and briefly discussed, including the conceptual design problem, the decomposition problem, the parameterization and optimization problem, and the large deformation problem. Notably, all these design problems are described using mathematical formulations in the framework of screw theory. This is further described in Fig. 4.6.

4.5   Conclusions

In conclusion, this chapter presents the construction of compliant mechanisms in both serial and parallel configurations. Particularly a mechanism-equivalent principle is utilized to guide the development of compliant mechanisms from single DOF flexible elements to multi DOF flexible elements. This leads to the identification of several common design problems, such as the conceptual and dimensional design, the construction and decomposition problem, as well as the force transmission in the large deformations. They are briefly discussed in Sect. 4.4 and will be further investigated in the following chapters in detail.

References

1. Joshi, S.A., Tsai, L.-W.: Jacobian analysis of limited-DOF parallel manipulators. In: ASME 2002 International Design Engineering Technical Conferences and Computers and Information in Engineering Conference, pp. 341–348. American Society of Mechanical Engineers (2002)
2. Dai, J.S.: Finite displacement screw operators with embedded chasles' motion. J. Mech. Robot. **4**(4), 041002 (2012)

Chapter 5
Conceptual Design of Compliant Parallel Mechanisms

5.1 Introduction

A compliant parallel mechanism is a type of compliant mechanisms [1] with a rigid moving platform directly connected to the base with more than one flexible elements [2]. Compared to other types of compliant mechanisms, compliant parallel mechanisms demonstrate higher stiffness, higher natural frequencies and higher precision, and have been successfully applied into the design of remote center-of-compliance (RCC) devices [3], force sensors [4, 5], ultra-precision positioning systems [6, 7], and nano-manipulators [8].

At the initial conceptual design stage, an important task in compliant parallel mechanism design is to seek a suitable arrangement of flexures and actuators for the desired mobility. Among the proposed approaches, the constraint-based approach [9, 10] has been widely used. In the constraint-based approach, flexible elements are represented with constraint lines and the constraint spaces are described in the forms of line-geometrical patterns. This is particularly suitable for flexible elements such as beam-type and blade-type flexures which have equal or more degrees of freedom than degrees of constraint.

However, the initial constraint-based design principle heavily relies on the designer's experience and the generated results are difficult to be generalized. Hopkins and Culpepper [11–13] improved the constraint-based approach and introduced the concept of actuation space in their proposed freedom and constraint topology approach [FACT], where actuation space was described in line patterns, and all freedom space, constraint space and actuation space were visualized by a set of geometric entities. Screw theory [14] was used to establish the relationship between constraint and actuation space, and a stiffness matrix was utilized to calculate the location and orientations of actuators [13].

Yu and Qiu [15, 16] simplified the synthesis process of actuation space using a screw-algebra based approach. This algebraic approach was inspired from the

© The Editor(s) (if applicable) and The Author(s), under exclusive license to Springer Nature Switzerland AG 2021
C. Qiu and J. S. Dai, *Analysis and Synthesis of Compliant Parallel Mechanisms—Screw Theory Approach*, Springer Tracts in Advanced Robotics 139,
https://doi.org/10.1007/978-3-030-48313-5_5

method proposed by Su and Tari [17], which revealed the constraint-based design approach utilize one special type of screw systems named the line screw system, and identified the relationship between the freedom space and constraint space as *reciprocal* [18]. In accordance with this, a synthesis criterion was proposed [16] to synthesize the actuation space of a compliant mechanism according to its linear independence with the constraint space. One advantage of this method is it helps designers obtain the initial design of actuator layouts quickly without the large input of geometrical dimensions and material parameters of a compliant platform.

To physically demonstrate the analytical design process of both constraint flexures and actuators, a compliant parallel mechanism employing shape-memory-alloy (SMA) based actuators is manufactured and evaluated with experiment tests. Among different types of actuators [19], shape-memory-alloy (SMA) based actuators are selected since they are suitable in the small-scale design of compliant parallel systems where space and load capacity are limited. They are widely used due to their unique pseudo-elastic features and shape recovery capabilities [20–23]. Compared with traditional piezoelectric actuators [24–26], SMA actuators are simple to design, lightweight, and are able to produce a large displacement per unit weight [27]. In this design, the SMA-spring can change one passive line constraint into an active line actuator. When the SMA spring is inactivated, the flexure behaves as a line constraint and resists the translation along its axis with bias springs [28]; when the SMA spring is heated with an electric current, it generates a translational stroke along the axis, thus enables the platform to change its mobility without rearranging the configurations of constraint and actuation spaces. Both FEA simulations and experiment observations are then conducted to examine the mobility of this platform, thus validating the conceptual design and the constraint-based design approach.

5.2   Conceptual Design of a Compliant Parallel Mechanism

Figure 5.1 presents the kinematic model of a compliant parallel mechanism when it is fully constrained with line constraint elements. Each line constraint element is composed of two spherical joints and one prismatic joint. Two bias springs (one is hidden in the intermediate block) functionalize as the constraints that compliantly resist the translation along its axis but resist no other form of motion. This is a common type of elastic elements used in designing elastically suspended system [29–31], which is available for further development and it is selected in designing this compliant platform.

Then in the framework of screw theory, the synthesis process of its constraint space and actuation space is briefly introduced, and two examples are given to demonstrate the synthesis procedure.

Fig. 5.1 Kinematic model
of the compliant parallel
mechanism

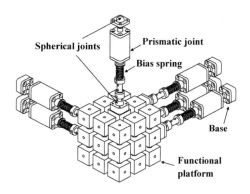

Spherical joints · Prismatic joint · Bias spring · Base · Functional platform

5.2.1 Synthesis of the Constraint Space

In screw theory, the freedom space, constraint space and actuation space can be represented in forms of screw systems [14]. A deflection occurring in a flexure system can be represented by a deformation twist if it is sufficiently small [29], which can be written as $t = [\theta^T \ \delta^T]^T$, where the primary part θ is the rotational deflection vector along the corresponding coordinate axes, and the second part δ is the translational deflection vector.

The effect of each line constraint can be represented with a line constraint wrench $w^r = [f^T \ m^T]^T$, where the primary part f is the force vector along the screw axes, and the second part m is the corresponding moment vector. The relationship between constraint wrench and deflection twist is named as *reciprocal* [18], which can be written as

$$(w^r)^T \Delta t = 0 \qquad (5.1)$$

where Δ is the elliptical polar operator [32] that interchanges the primary and second parts of a screw vector, and its formula has been given in Eq. (2.7). The initial constraint space can be obtained directly from the null space of freedom space according to Eq. (5.1). Since the function of each line constraint works as a pure force, additional conditions [17] should be added as

$$f \cdot m = 0, \ \|f\| \neq 0 \qquad (5.2)$$

In order to obtain line constraint wrenches, initial constraint wrenches should be modified according to the additional criteria in Eq. (5.2), and they can be aggregated into the required line constraint space. For example, when the compliant parallel mechanism is fully constrained, the freedom space is $\mathbb{S} = \{\varnothing\}$. The constraint space $\mathbb{S}^r$ can be calculated according to Eq. (5.1), where the initial solution can be easily determined as

Fig. 5.2 Fully constraint
configuration

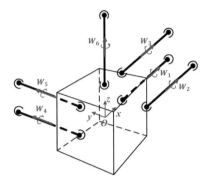

$$\mathbb{S}^r = \begin{cases} \boldsymbol{w}_1^r = \begin{bmatrix} 1 & 0 & 0 & 0 & 0 & 0 \end{bmatrix}^\mathrm{T} \\ \boldsymbol{w}_2^r = \begin{bmatrix} 0 & 1 & 0 & 0 & 0 & 0 \end{bmatrix}^\mathrm{T} \\ \boldsymbol{w}_3^r = \begin{bmatrix} 0 & 0 & 1 & 0 & 0 & 0 \end{bmatrix}^\mathrm{T} \\ \boldsymbol{w}_4^r = \begin{bmatrix} 0 & 0 & 0 & 1 & 0 & 0 \end{bmatrix}^\mathrm{T} \\ \boldsymbol{w}_5^r = \begin{bmatrix} 0 & 0 & 0 & 0 & 1 & 0 \end{bmatrix}^\mathrm{T} \\ \boldsymbol{w}_6^r = \begin{bmatrix} 0 & 0 & 0 & 0 & 0 & 1 \end{bmatrix}^\mathrm{T} \end{cases} \qquad (5.3)$$

However, not all constraint wrenches obtained in Eq. (5.3) satisfy the criteria in Eq. (5.2), they can be modified and one possible solution is

$$\mathbb{S}^r = \begin{cases} \boldsymbol{w}_1^r = \begin{bmatrix} 1 & 0 & 0 & 0 & 0 & 0 \end{bmatrix}^\mathrm{T} \\ \boldsymbol{w}_2^r = \begin{bmatrix} 1 & 0 & 0 & 0 & 0 & 1 \end{bmatrix}^\mathrm{T} \\ \boldsymbol{w}_3^r = \begin{bmatrix} 1 & 0 & 0 & 0 & 1 & 0 \end{bmatrix}^\mathrm{T} \\ \boldsymbol{w}_4^r = \begin{bmatrix} 0 & 1 & 0 & 0 & 0 & 0 \end{bmatrix}^\mathrm{T} \\ \boldsymbol{w}_5^r = \begin{bmatrix} 0 & 1 & 0 & 1 & 0 & 0 \end{bmatrix}^\mathrm{T} \\ \boldsymbol{w}_6^r = \begin{bmatrix} 0 & 0 & 1 & 0 & 0 & 0 \end{bmatrix}^\mathrm{T} \end{cases} \qquad (5.4)$$

The corresponding configuration is also presented in Fig. 5.2.

5.2.2   Synthesis of the Actuation Space

The actuation system of compliant parallel platform is calculated in a similar way. In accordance with the line constraint definition, only linear actuator that generates translational displacement along its axis is considered, which can be represented as a line actuation wrench $\boldsymbol{w}^a = [\boldsymbol{f}^\mathrm{T} \quad \boldsymbol{m}^\mathrm{T}]^\mathrm{T}$ [13]. Similar to the definition of constraint wrench, additional conditions should be added with the actuation wrench as $\boldsymbol{f} \cdot \boldsymbol{m} = 0$ and $\|\boldsymbol{f}\| \neq 0$. The line actuation wrench can be obtained simply by swapping the rotational vector and translational vector of deflection twist [16], which can be written as

$$\boldsymbol{w}^a = \boldsymbol{t}\Delta \tag{5.5}$$

However not all actuation wrenches calculated based on Eq. (5.5) are line actuation wrenches, and they have to be modified according to the criterion that *actuation wrench is linearly independent with constraint wrench* [16]. For example, assume the compliant platform has 3 DOF $R_x R_y R_z$, thus the freedom space can be written as

$$\mathbb{S} = \begin{cases} \boldsymbol{t}_1 = \begin{bmatrix} 1 & 0 & 0 & 0 & 0 & 0 \end{bmatrix}^{\mathrm{T}} \\ \boldsymbol{t}_2 = \begin{bmatrix} 0 & 1 & 0 & 0 & 0 & 0 \end{bmatrix}^{\mathrm{T}} \\ \boldsymbol{t}_3 = \begin{bmatrix} 0 & 0 & 1 & 0 & 0 & 0 \end{bmatrix}^{\mathrm{T}} \end{cases} \tag{5.6}$$

Then the constraint space of this configuration can be determined according to Eq. (5.1) as

$$\mathbb{S}^r = \begin{cases} \boldsymbol{w}_1^r = \begin{bmatrix} 1 & 0 & 0 & 0 & 0 & 0 \end{bmatrix}^{\mathrm{T}} \\ \boldsymbol{w}_2^r = \begin{bmatrix} 0 & 1 & 0 & 0 & 0 & 0 \end{bmatrix}^{\mathrm{T}} \\ \boldsymbol{w}_3^r = \begin{bmatrix} 0 & 0 & 1 & 0 & 0 & 0 \end{bmatrix}^{\mathrm{T}} \end{cases} \tag{5.7}$$

Finally, the actuation space is calculated according to Eq. (5.5), which can be calculated as

$$\mathbb{S}^a = \begin{cases} \boldsymbol{w}_1^a = \begin{bmatrix} 0 & 0 & 0 & 1 & 0 & 0 \end{bmatrix}^{\mathrm{T}} \\ \boldsymbol{w}_2^a = \begin{bmatrix} 0 & 0 & 0 & 0 & 1 & 0 \end{bmatrix}^{\mathrm{T}} \\ \boldsymbol{w}_3^a = \begin{bmatrix} 0 & 0 & 0 & 0 & 0 & 1 \end{bmatrix}^{\mathrm{T}} \end{cases} \tag{5.8}$$

However, the initial selection of actuation wrenches in Eq. (5.8) does not satisfy the force actuator requirements, and they can be modified similarly to that of constraint wrenches as

$$\mathbb{S}^a = \begin{cases} \boldsymbol{w}_1^a = \begin{bmatrix} 1 & 0 & 0 & 0 & 0 & 1 \end{bmatrix}^{\mathrm{T}} \\ \boldsymbol{w}_2^a = \begin{bmatrix} 1 & 0 & 0 & 0 & 1 & 0 \end{bmatrix}^{\mathrm{T}} \\ \boldsymbol{w}_3^a = \begin{bmatrix} 0 & 1 & 0 & 1 & 0 & 0 \end{bmatrix}^{\mathrm{T}} \end{cases} \tag{5.9}$$

The schematic diagrams of the constraint and actuation space of this configuration are shown in Fig. 5.3. Actually Figs. 5.2 and 5.3 present the same configuration of compliant parallel mechanism combining the constraint and actuation wrenches, the difference comes from the number and arrangement of the actuation system.

According to the two fundamental equations (5.1) and (5.5), as well as their corresponding modification processes, both line constraint space and line actuation space of desired motion patterns can be developed. One program is developed in

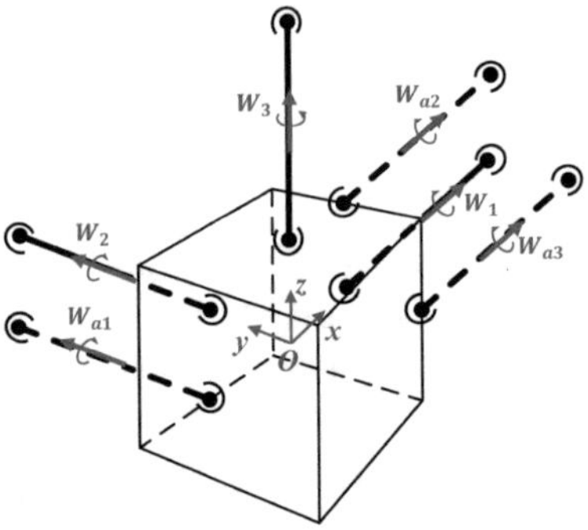

Fig. 5.3 Configuration with
3DOF $R_x R_y R_z$

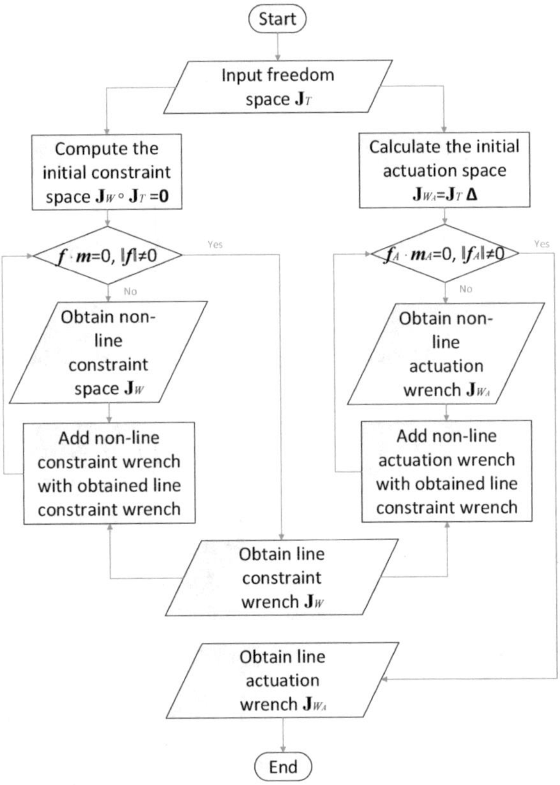

Fig. 5.4 Flowchart of
actuation space synthesis
process based on the Matlab
program

the Matlab software to calculate the constraint and actuation space with respect to
various degrees of freedom of the compliant parallel mechanism, and the flow chart
is shown in Fig. 5.4.

Further Appendix A provides the table of constraint space and actuation space
corresponding to the mobilities from 1 DOF R_x to 4 DOF $R_x R_y R_z P_z$. In order to
change the mobility of the platform without rearranging the constraint elements, the

passive constraint wrench should be able to be transformed into an active actuation wrench, this can be realized with shape-memory-alloy based actuation systems [21, 33]. The SMA coil spring behaves as a passive spring when it is unheated (in the martensitic phase) and generate deflection along its axis when it is heated (in the austenitic phase). Combined with bias springs, this SMA-spring actuator can work both as a passive line constraint and an active line actuator, thus enable the compliant parallel mechanism to have maximum 4 degrees of freedom. In the following Sect. 5.3, Detailed Structure design of this SMA-spring embedded linear actuator is provided, including experiments regarding the force-deflection characteristics of this actuator.

5.3 Design of SMA-Spring Based Linear Actuator

The preferred embodiment of SMA-spring linear actuator is illustrated in Fig. 5.5. The function of a traditional linear constraint element is to compliantly resist the translation along its axis but resist no other form of motion [28]. It consists of two spherical joints 1 and 8, and one prismatic joint 5. The inner end of spherical joint 1 is coupled with shaft 2. Prismatic joint 5 is coupled with spherical joint 8, and it can move along and rotate around shaft 2 freely. Two bias springs 6 unilaterally constrain prismatic joint 5 so that an elastic constraint force can be generated to resist the translation of prismatic joint 5. One bias spring is assembled between slider block 3 and prismatic joint 5, and the other one is assembled between prismatic joint 5 and slider block 7 that is attached at the end of shaft 2. Slider block 3 can change its location on shaft 2, so the preloads of two bias springs can be predetermined to eliminate the translational backlash, also the effective stiffness of constraint along the shaft axis can be designed.

Subsequently, SMA springs are embedded into the structure to demonstrate the feasibility of one DOF linear actuators. As shown in Fig. 5.5, two SMA springs 4 are attached between slider block 3 and prismatic joint 5. The SMA spring is activated by electrical current. When the SMA spring is activated, the coils are shrank and prismatic joint 5 is pulled towards slider block 3, and the whole structure reaches its balance at a new position, thus generates a deflection along the shaft axis. When the

Fig. 5.5 Linear actuator, components include joints 1 and 8, intermediate shaft 2, slider block 3 and 7, SMA springs 4, prismatic joint 5, bias springs 6

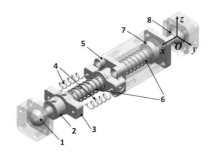

SMA spring is not activated, the prismatic joint will return to its original position with the balanced force from compressed bias springs.

Based on the structure of the SMA linear actuator, it is important to select suitable types of bias springs and SMA springs to determine the stroke of the actuator. Firstly analytical model and experiment tests of bias springs are provided, and suitable ones are selected according to the dimensions of line constraint structure. Then the two-state force-deflection behaviours of SMA springs are studied. Their properties are obtained from experiment tests, based on what analytical models are developed, and they are used in determining the suitable SMA spring dimensions in accordance with the stiffness of bias springs. Finally, their joint performances are demonstrated with experiment tests.

5.3.1 Selection of Bias Springs

As described in Sect. 5.3, two bias springs are employed to provide resistance force along the axis of the line constraint element. Each bias spring is an alloy steel helical compression spring with spring diameter $D = 5.35 \times 10^{-3}$ m, wire diameter $d = 500 \times 10^{-6}$ m, spring length $L = 17 \times 10^{-3}$ m and active coils $n = 8$, which are determined according to the geometrical dimensions of line constraint structure. Shear modulus of each bias spring is $G = 79.3 \times 10^{9}$ Pa [34]. In a typical steel coil spring, the relationship between axial force F and deflection δ can be described as a simple linear equation [35]

$$F = \frac{Gd^4}{8D^3n}\delta \qquad (5.10)$$

The resulted stiffness of each bias spring is $k = 0.506$ N/mm according to Eq. (5.10). In the line constraint element, two bias springs are attached serially. By adjusting the position of slider block 3 in Fig. 5.5, each spring is pre-compressed 5 mm, thus the stiffness of line constraint equals to $K_t = 2k = 1.012$ N/mm within the deflection range of 5 mm.

Further force-displacement relationship of the line constraint structure is tested with experiments. The experiment setup with a line constraint element assembled is illustrated in Fig. 5.6a. The experiment setup includes a linear guide that is driven by a brushless motor, and it can generate required deflection of bias springs along the axis of the line constraint structure. An ATI nano 17 force/torque sensor is coupled with the moving platform of linear guide and records the axial resisted force of bias springs. The force-deflection relationship of the bias-spring system is provided in Fig. 5.6b, which shows both the loading and unloading curve are almost overlapping each other during the designed deflection range, and the slopes of them remain linear as well. The experimental results of constraint stiffness $K_e = 1.00$ N/mm, which is quite close to the theoretical result of K_t.

5.3.2 Selection of SMA Springs

Following the study of stiffness characteristics of bias springs in the line constraint structure, the force-deflection characteristics of SMA springs are explored in this section. The static two-state model of SMA coil spring actuator proposed by [33] is selected which is able to describe the nonlinear force-deflection feature in Martensitic 100% state. Following the design framework of SMA coil spring actuator, experiments of SMA coil spring samples were conducted firstly to obtain properties of SMA spring actuator. Based on the obtained parameters, analytical models were developed to verify the experimental tests of selected samples. Then the analytical models were used to determine suitable SMA springs together with stiffness characteristics of bias springs obtained in Sect. 5.3.1.

5.3.2.1 Obtaining Properties of SMA Springs

The properties of the SMA springs were obtained from tensile tests of SMA coil spring specimens. The tensile testing experiment setup is the same as the one that tests bias springs in Fig. 5.6a. SMA coil springs were made by wounding commercial SMA wires (Flexinol by Dynalloy, CA, USA) around a steel rod, and they were annealed at a temperature of 500 °C for 15 min to memory the coil spring shape. The geometric parameters of the tested SMA coil springs are including wire diameter $d = 0.375$ mm, spring diameter $D = 3.25$ mm, active coil turns are 5.

The properties needed to be obtained from tensile tests of SMA coil springs include the shear modulus G_A in austenite phase, the shear modulus G_M in martensite phase, the critical stress τ_A^{cr} for the maximum shear strain (1%) in austenite phase, the critical stresses τ_s^{cr} and τ_f^{cr} for the beginning (1%) and end (6%) of stress induced detwinning process in martensite phase. The force-deflection characteristics of tested SMA coil springs were obtained directly from tensile test experiments. In detail, the tensile

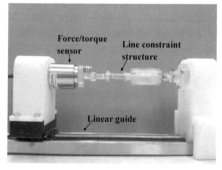

(a) Experiment setup

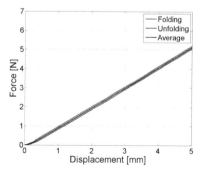

(b) Force-deflection characteristics

Fig. 5.6 Experiment setup and resulted force-deflection characteristics of bias springs

Table 5.1 Properties of SMA coil springs

$A_{100\%}$ state	G_A (GPa)	τ_A^{cr} (MPa)		
	21	230		
$M_{100\%}$ state	G_M (GPa)	τ_s^{cr} (MPa)	τ_f^{cr} (MPa)	γ_L
	8.27	76	150	0.05

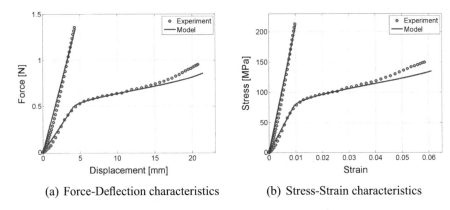

(a) Force-Deflection characteristics   (b) Stress-Strain characteristics

Fig. 5.7 Force-Deflection and Stress-Strain characteristics of tested SMA springs

tests of specimens in the martensite phase were conducted at 23 °C in temperature and 50% relative humidity. The experiments of specimens in the austenite phase were conducted by adding a voltage of 0.9 V that heats the SMA spring. The deflection value and force value were converted into shear strain value and shear stress value using formulae which can be found in [33], thus both force-deflection curves and stress-strain curves were obtained, and the properties of SMA spring specimens were calculated accordingly, which were listed in Table 5.1.

Analytical models of SMA springs were developed from the physical properties, and they are compared with experimental results in Fig. 5.7, where Fig. 5.7a illustrates the comparisons of force-deflection characteristics, and Fig. 5.7b compares the stress-strain characteristics. The results show close agreements between experimental results and analytical models, and the discrepancy appears with the increasing deformation of specimens that is due to the spring curvature effect and the linear stress distribution assumption used in the equations [33].

5.3.2.2   Selection of SMA Springs

Following the identifications of SMA spring properties, the design of SMA-spring linear actuator is accomplished in this section. Generally, the maximum stroke and maximum force are pre-determined by designers according to the geometrical constraints and stiffness characteristics of the actuation systems, in this paper, the stroke

Fig. 5.8 Preselection of
SMA spring with bias spring

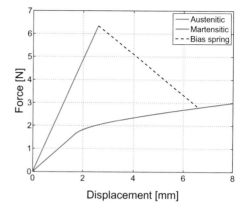

Table 5.2 Geometric parameters of the designed SMA coil springs

Spring index	Wire diameter (μm)	Spring diameter (mm)	No. of turns
C	d (μm)	D (mm)	n
6	500	3.00	5

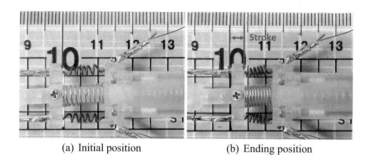

(a) Initial position   (b) Ending position

Fig. 5.9 Initial and ending positions of SMA-spring linear actuator

is selected to be 4 mm according to the geometrical constraints of line constraint
structure and the stiffness characteristics of bias springs studied in Sect. 5.3.1. The
result is illustrated in Fig. 5.8 with the properties of SMA spring provided in Table 5.2.

Following the analytical calculation of stroke of the SMA actuator, the experiment
was conducted in comparison with the analytical results. Two SMA springs with
properties obtained above were manufactured and assembled in parallel with the two
bias springs. The experiment results are illustrated in Fig. 5.9, from which we can
obtain the actual stroke of the linear actuator which is about 4 mm. This result is
quite close to the prediction of analytical models. Based on the experiment results
of the SMA actuator, in the following sections, the mobility and workspace of the
compliant parallel mechanism are examined with both finite-element-simulation and
experiment observation in Sect. 5.4.

5.4   FEA Simulation and Experiment Validation

In Sect. 5.3, both force-deflection characteristics of passive bias springs and active SMA coil spring actuators are explored and proper linear actuators are developed based on the combinations of bias springs and SMA springs. According to the experiment results of the two-state force-deflection performance of single SMA linear actuator, it is able to verify the mobility and workspace of the compliant parallel mechanism developed in Sect. 5.2. Both Finite-Element-Analysis simulation and experiment observation are used to validate the behaviours of the platform with various types of degree-of-freedom.

5.4.1   FEA Simulation

FEA simulation is conducted firstly in this section. The workspace of the parallel mechanism is determined by the strokes of activated SMA spring actuators, which are based on the deformations of SMA springs and bias springs, and all other components behave as rigid parts. For that reason, a rigid transient analysis for the multi-body system is set up using commercial ANSYS software. In each line constraint structure, spherical joints and prismatic joints are modelled by imposing appropriate kinematic constraints. Translational spring elements are imported between the slider block and prismatic joint to simulate the effect of bias springs. The free lengths of spring elements are set to be 17 mm, and they are preloaded with 5 mm compressions. For each bias spring, a stiffness of 500 N/m is added. Further the effects of SMA springs are simulated by adding displacement loads on the linear actuator element. The displacement load is imported as time-displacement data, and the maximum stroke is set to be 4 mm according to the experimental tests conducted in Sect. 5.3.2.

Following the settings of FEA simulation, Mobilities of the parallel mechanism are examined. Since all types of DOF are the combinations of four one degree-of-freedom motions, including translation P_z along z axis, and rotations R_x, R_y, R_z about each axis, and they are selected in the present work. The positions and orientations of actuators are determined according to the corresponding actuation space listed in Appendix A. The selected simulation results of DOF R_x and P_z are presented in Fig. 5.10a, c, from which we can see the parallel compliant mechanism clearly demonstrates the required degrees of freedom with a large working range.

5.4.2   Physical Prototype Demonstration

Following the FEA simulation of parallel compliant mechanism, experiment tests of the physical prototype are conducted in this section. The physical prototype contains three parts, including the functional platform, the base and the compliant support-

Fig. 5.10 FEA simulation and experiment observation of selected prototype motions

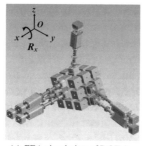
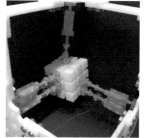

(a) FEA simulation of DOF R_x (b) Experiment result of DOF R_x

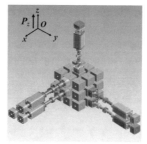
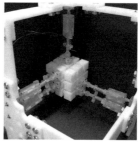

(c) FEA simulation of DOF P_z (d) Experiment result of DOF P_z

ing elements. All the components are made of plastic using rapid prototyping 3-D printers, they are strong and light weight. The compliant supporting elements are composed of line constraints and SMA-spring linear actuators. The selected testing results are illustrated in Fig. 5.10b, d, where Fig. 5.10b shows the deformation of functional platform with DOF R_x, and Fig. 5.10d illustrates the deformation of platform with DOF P_z. They are also compared with corresponding FEA simulation results. These two comparison results validate FEA simulations and demonstrate the large workspace of the functional platform. In future work, proper measurement systems will be used to measure the motions of the parallel mechanism in comparisons with FEA simulation calculation results.

5.5 Conclusions

This chapter presented a comprehensive study of the conceptual design of compliant mechanisms and developed a novel compliant parallel mechanism using SMA-spring actuators. The design of this parallel mechanism was inspired from the conventional elastically suspended systems, and it creatively employed shape-memory-alloy coil springs that can actively transform traditional line constraint elements into linear actuators, which enabled the functional platform to obtain various degrees of freedom according to the numbers and positions of SMA-spring actuators. Constraint-based approach and screw theory were utilized to synthesize the configuration of the compliant parallel mechanism, where the constraints and actuators are represented with line geometries in constraint-based approach, and they are mathematically described as line screw vectors in the framework of screw theory. Subsequently, the constraint space and actuation space with various degrees of freedoms from 1 DOF to 4 DOF were obtained according to the freedom space of parallel platform.

Following the synthesizing process of the actuation space of compliant parallel mechanisms, SMA-spring based linear actuator was developed. Two important segments including bias springs and SMA springs were studied and tested to obtain their properties. Based on their joint force-deflection characteristics, linear actuators with the desired stroke were developed, and they were applied to the design of a physical prototype. Both finite element simulations and experimental observations of the prototype undertaking various motions were conducted, and the results demonstrated a large workspace of each degree of freedom. This result suggested the potential value of SMA actuation systems in developing compliant mechanisms.

References

1. Howell, L.L.: Compliant Mechanisms. Wiley-Interscience (2001)
2. Smith, S.T.: Flexures: Elements of Elastic Mechanisms. CRC Press (2000)
3. Drake, S.H.: Using compliance in lieu of sensory feedback for automatic assembly. Ph.D. thesis, Massachusetts Institute of Technology (1978)
4. Dai, J.S., Kerr, D.: A six-component contact force measurement device based on the Stewart platform. Proc. Inst. Mech. Eng. Part C: J. Mech. Eng. Sci. **214**(5), 687–697 (2000)
5. Ataollahi, A., Fallah, A.S., Seneviratne, L.D., Dasgupta, P., Althoefer, K.: Novel force sensing approach employing prismatic-tip optical fiber inside an orthoplanar spring structure
6. Awtar, S., Slocum, A.H.: Constraint-based design of parallel kinematic XY flexure mechanisms. J. Mech. Des. **129**(8), 816–830 (2007)
7. Kim, H.-Y., Ahn, D.-H., Gweon, D.-G.: Development of a novel 3-degrees of freedom flexure based positioning system. Rev. Sci. Instrum. **83**(5), 055114–055114 (2012)
8. Jensen, K.A., Lusk, C.P., Howell, L.L.: An XYZ micromanipulator with three translational degrees of freedom. Robotica **24**(3), 305–314 (2006)
9. Blanding, D.L.: Exact constraint: machine design using kinematic processing. American Society of Mechanical Engineers (1999)
10. Hale, L.C.: Principles and techniques for designing precision machines. Technical report, Lawrence Livermore National Lab., CA, USA (1999)
11. Hopkins, J.B.: Design of flexure-based motion stages for mechatronic systems via freedom, actuation and constraint topologies (FACT). Ph.D. thesis, Massachusetts Institute of Technology (2010)
12. Hopkins, J.B., Culpepper, M.L.: Synthesis of multi-degree of freedom, parallel flexure system concepts via freedom and constraint topology (fact)-part I: Principles. Precis. Eng. **34**(2), 259–270 (2010)
13. Hopkins, J.B., Culpepper, M.L.: A screw theory basis for quantitative and graphical design tools that define layout of actuators to minimize parasitic errors in parallel flexure systems. Precis. Eng. **34**(4), 767–776 (2010)
14. Ball, R.S.: A Treatise on the Theory of Screws. Cambridge University Press (1900)
15. Qiu, C., Yu, J., Li, S., Su, H.J., Zeng, Y.: Synthesis of actuation spaces of multi-axis parallel flexure mechanisms based on screw theory. In: ASME 2011 International Design Engineering

Technical Conferences and Computers and Information in Engineering Conference, pp. 181–190. American Society of Mechanical Engineers (2011)

16. Yu, J.J., Li, S.Z., Qiu, C.: An analytical approach for synthesizing line actuation spaces of parallel flexure mechanisms. J. Mech. Des. **135**(12), 124501–124501 (2013)
17. Su, H.-J., Tari, H.: Realizing orthogonal motions with wire flexures connected in parallel. J. Mech. Des. **132**, 121002 (2010)
18. Dai, J.S., Rees Jones, J.: Interrelationship between screw systems and corresponding reciprocal systems and applications. Mech. Mach. Theory **36**(5), 633–651 (2001)
19. Huber, J., Fleck, N., Ashby, M.: The selection of mechanical actuators based on performance indices. Proc. Royal Soc. Lond. Ser. A: Math. Phys. Eng. Sci. **453**(1965), 2185–2205 (1997)
20. Morgan, N.: Medical shape memory alloy application-the market and its products. Mater. Sci. Eng. A **378**(1), 16–23 (2004)
21. Kim, B., Lee, M.G., Lee, Y.P., Kim, Y., Lee, G.: An earthworm-like micro robot using shape memory alloy actuator. Sens. Actuators A **125**(2), 429–437 (2006)
22. Hartl, D., Lagoudas, D.C.: Aerospace applications of shape memory alloys. Proc. Inst. Mech. Eng. Part G: J. Aerosp. Eng. **221**(4), 535–552 (2007)
23. Kyung, J., Ko, B., Ha, Y., Chung, G.: Design of a microgripper for micromanipulation of microcomponents using SMA wires and flexible hinges. Sens. Actuators A **141**(1), 144–150 (2008)
24. Li, Y., Xu, Q.: A novel piezoactuated XY stage with parallel, decoupled, and stacked flexure structure for micro-/nanopositioning. IEEE Trans. Ind. Electron. **58**(8), 3601–3615 (2011)
25. Kang, D., Gweon, D.: Development of flexure based 6-degrees of freedom parallel nano-positioning system with large displacement. Rev. Sci. Instrum. **83**(3), 035003–035003 (2012)
26. Yong, Y., Moheimani, S., Kenton, B., Leang, K.: Invited review article: high-speed flexure-guided nanopositioning: mechanical design and control issues. Rev. Sci. Instrum. **83**(12), 121101–121101 (2012)
27. Van Humbeeck, J.: Non-medical applications of shape memory alloys. Mater. Sci. Eng. A **273**, 134–148 (1999)
28. Schimmels, J.M., Huang, S.: Spatial parallel compliant mechanism, Feb 2000. US Patent 6,021,579
29. Dimentberg, F.M.: The screw calculus and its applications in mechanics. Technical report, DTIC Document (1968)
30. Patterson, T., Lipkin, H.: A classification of robot compliance. Trans. Am. Soc. Mech. Eng. J. Mech. Des. **115**, 581–581 (1993)
31. Huang, S., Schimmels, J.M.: The eigenscrew decomposition of spatial stiffness matrices. IEEE Trans. Robot. Autom. **16**(2), 146–156 (2000)
32. Lipkin, H., Duffy, J.: The elliptic polarity of screws. ASME J. Mech. Trans. Autom. Des. **107**, 377–387 (1985)
33. An, S.-M., Ryu, J., Cho, M., Cho, K.-J.: Engineering design framework for a shape memory alloy coil spring actuator using a static two-state model. Smart Mater. Struct. **21**(5), 055009 (2012)
34. Institute, S.M.: Handbook of Spring Design. Spring Manufacturers Institute Inc. (1991)
35. Shigley, J.E., Mischke, C.R., Budynas, R.G., Liu, X., Gao, Z.: Mechanical Engineering Design, vol. 89. McGraw-Hill, New York (1989)

Chapter 6
Stiffness Construction and Decomposition of Compliant Parallel Mechanisms

6.1 Introduction

Initial conceptual design of compliant mechanisms is accomplished in Chap. 5, from which we can obtain the layout of flexible elements according to required degrees of freedom. This can lead to further stiffness performance evaluation of the developed compliant parallel platform. Two types of stiffness problems exist, including the stiffness analysis and its synthesis. In the stiffness analysis, limbs are modelled as elastic elements, which are used to construct the platform's stiffness matrix. Based on the developed stiffness matrix, subsequent studies can be carried out, such as the stiffness optimization design [1] and sensitivity analysis [2], etc. Common approaches in determining the layouts of elastic elements include type synthesis approach [3], constraint-based approach [4] and screw-algebra based synthesis approaches [5–7]. The fundamental principle is to find out the configuration of elastic elements according to the relationship between motions and wrenches [8, 9].

Different from stiffness analysis which constructs the stiffness matrix based on the known configuration of a parallel platform, stiffness synthesis aims to decompose the pre-determined stiffness inversely and obtain the corresponding layout of elastic elements. Stiffness synthesis is particularly useful when researchers need to design a parallel platform for a given stiffness performance when there is no obvious solution. Though it is important, the study of stiffness synthesis has a relatively short history. In the early stages, researches have been focusing on identifying the properties of stiffness matrix needed to conduct stiffness synthesis, most of which originated from the study of elastically suspended platforms [10–12]. Ball [10] first introduced the screw theory and applied it to the study of their rigid body motions. Dimentberg [11] used screw theory to study the static and small vibrations of a rigid body elastically suspended by line springs. Loncaric [12] analyzed the spatial compliance/stiffness matrix using Lie groups. He first showed that only stiffnesses with zero trace off-diagonal matrices are realizable by line springs.

C. Qiu and J. S. Dai, *Analysis and Synthesis of Compliant Parallel Mechanisms—Screw Theory Approach*, Springer Tracts in Advanced Robotics 139, https://doi.org/10.1007/978-3-030-48313-5_6

Loncaric's realization theorem led to the further development of stiffness synthesis approaches [13–17]. Huang and Schimmels [13] identified the Loncaric's realization theorem as the isotropic condition and used it to decompose the spatial stiffness with simple springs. The proposed algorithm was based on the Cholesky decomposition and synthesized the stiffness matrix with no more than seven simple springs. Ciblak and Lipkin [14] and Roberts [15] utilized the isotropic condition to synthesize the spatial stiffness with a minimal number of simple springs that is equal to the rank of a stiffness matrix. Dai and Ding [16, 17] used adjoint transformation to construct the isotropic compliance matrix of a parallel platform and evaluated the effect of design parameters on eigen-compliances [18]. In their approaches, since the simple springs are decomposed recursively from the stiffness matrix, they are named as direct-recursion algorithms.

Recently more algorithms [19–22] have been proposed in order to generate meaningful synthesis results, such as layouts of elastic elements with either realizable geometrical arrangements [19] or realizable stiffness characteristics [21]. Ciblak and Lipkin [19] proposed a decomposition method that considers the geometrical properties of decomposed springs, inspired by the eigenvalue decomposition proposed by Lipkin and Patterson [23]. Other approaches [20, 21] are also proposed to consider the properties of synthesized springs. Unlike the direct-recursion algorithm, these methods partition the stiffness matrix into sub-matrices and decompose each of them separately, thus they are identified as the matrix-partition algorithm. Algorithms regarding stiffness matrices that don't satisfy the isotropic condition are proposed in [24–26]; the stiffness matrix of loaded conditions are investigated in [27, 28]. Compared to the direct-recursion algorithm, the matrix-partition algorithm can determine properties of synthesized results, though the selection of them require a better understanding of the stiffness matrix to be decomposed.

Despite the fact both direct-recursion and matrix-partition algorithms have been proposed, they are rarely utilized to synthesize a stiffness matrix developed in the analysis process, making it difficult to evaluate their effectiveness from a practical-design point of view. To solve this problem requires an understanding of both stiffness construction and decomposition, particularly the intrinsic relationship between them. As a consequence, the compliant parallel mechanisms developed in Chap. 5 are utilized to evaluate the proposed algorithms and explore the relationship between stiffness construction and decomposition process. As a result, both stiffness analysis and synthesis of the constraint stiffness of the developed compliant parallel platforms are conducted in this chapter. In Sect. 6.2, the layout design of constraint limbs are described in detail, which is used to construct the constraint stiffness matrix. Then the existing synthesis algorithms are compared in Sect. 6.3. Particularly, a detailed exploration of the matrix-partition algorithm establishes a one-to-one correspondence between the synthesized result and the configuration of constraint limbs used to construct the compliant parallel platform. This is further verified with several examples presented in Sect. 6.4.

6.2 Constraint Stiffness Construction of the Compliant Parallel Platform

The compliant parallel mechanism [29] is shown in Fig. 6.1, whose initial aim is to demonstrate the relationship between motions and constraints. This parallel mechanism contains a fixed base and a functional platform, they are linked with six compliant limbs. One limb is an active linear actuator using shape-memory-alloy (SMA) springs, the other five are passive constraint limbs. Each constraint limb consists of two spherical joints at its two ends and one prismatic joint along its intermediate shaft. Its function is to compliantly resist the translation along its axis but no other forms of motion. Two bias springs are embedded along its axis to provide resisting stiffness. The SMA actuator has the same structure as the passive constraint limb but adds additional shape-memory-alloy springs. When the SMA springs are activated by electrical current, they shrink and generate a stroke along the shaft axis; when the SMA springs are not activated, the line constraint structure is pushed back to its original length by the bias springs [29].

6.2.1 Construct the Constraint Stiffness Matrix

To construct the constraint stiffness matrix, we need to know the layout of constraint limbs, which can be designed according to the relationship between motions and constraints. The synthesis of constraint limbs according to the reciprocal relationship with motions of the platform has been given in Sect. 5.2.1 in Chap. 5 and they are omitted here. Following the layout design of constraint limbs, the corresponding stiffness matrix is constructed. The stiffness matrices of various types of compliant parallel platforms have been evaluated in [16, 30, 31], where a common approach

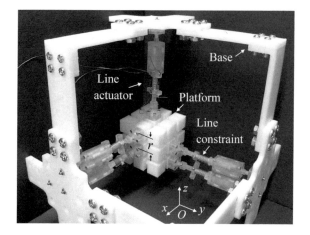

Fig. 6.1 The SPS-orthogonal parallel mechanism with one SMA-linear actuator and five line constraints

is to construct the platform's stiffness matrix by integrating each limb's stiffness matrix. One constraint limb of the compliant parallel mechanism used in this chapter is shown in Fig. 6.2, which includes two spherical joints at both of its ends and one prismatic joint in the middle. Two bias springs are assembled along with the prismatic joint so that it can resist forces along both two directions of the axis. According to the mechanical functionality of the constraint limb, its stiffness matrix can be presented as

$$\mathbf{K}_i = k_i \mathbf{S}_i \mathbf{S}_i^{\mathrm{T}} \tag{6.1}$$

where $\mathbf{S}_i$ is the screw associated with one constraint limb and $\mathbf{S}_i = \left[\mathbf{n}_i^{\mathrm{T}} \quad (\mathbf{r}_i \times \mathbf{n}_i)^{\mathrm{T}} \right]^{\mathrm{T}}$. $\mathbf{n}_i$ is a 3×1 unit vector along the axial axis of constraint i, $\mathbf{r}_i$ is the position vector of $\mathbf{n}_i$. k_i is the stiffness coefficient of one constraint which is the sum of two bias springs' stiffnesses. $\boldsymbol{w}_i$ in Eq. 6.8 is the exerted wrench by constraint limb and it can be written as $f \mathbf{S}_i$. The coordinate frame used to describe $\boldsymbol{w}_i$ is chosen to be the global coordinate frame $\{O, x, y, z\}$ shown in Fig. 6.3. Further it can be verified that

$$\mathbf{S}_i^{\mathrm{T}} \Delta \mathbf{S}_i = 0 \tag{6.2}$$

which indicates the constraint limb functionalize as a simple linear spring. The constraint stiffness matrix can be developed from Eq. (6.1) as well as the layouts of constraint limbs provided in Sect. 6.2.2, which has the form

$$\mathbf{K} = \sum_{i=1}^{r} k_i \mathbf{S}_i \mathbf{S}_i^{\mathrm{T}} \tag{6.3}$$

where $n = rank\{\mathbb{S}^c\}$ and it is the number of the constraint limbs. Substituting Eq. (6.2) into Eq. (6.3) and using the linear mapping property of trace operation, it follows that

Fig. 6.2 Physical prototype of one constraint limb

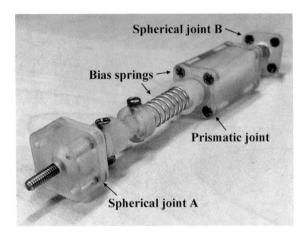

$$tr(\mathbf{K}\Delta) = tr\left(\sum_{i=1}^{r} k_i \mathbf{S}_i \mathbf{S}_i^{\mathrm{T}} \Delta\right) = \sum_{i=1}^{r} tr\left(k_i \mathbf{S}_i \mathbf{S}_i^{\mathrm{T}} \Delta\right) \tag{6.4}$$

Further Eq. (6.4) can be simplified using the property of the product's trace operation, which can be written as

$$tr(\mathbf{K}\Delta) = \sum_{i=1}^{n} k_i tr\left\{\mathbf{S}_i(\Delta\mathbf{S}_i)^{\mathrm{T}}\right\} = \sum_{i=1}^{n} k_i tr(\mathbf{S}_i^{\mathrm{T}} \Delta\mathbf{S}_i)$$
$$= \sum_{i=1}^{n} k_i(\mathbf{S}_i^{\mathrm{T}} \Delta\mathbf{S}_i) = 0 \tag{6.5}$$

This shows the constraint stiffness matrix has off-diagonal sub-matrices that are traceless. This is also identified as the *isotropic* condition [13]. When the mobility of the parallel platform changes, the number of linearly independent constraints also change. As a result, the obtained constraint stiffness matrix generally has a rank less than six and are positive semi-definite.

6.2.2 An Example of Constraint Stiffness Construction

An example is provided in this section to demonstrate the layout design of constraint limbs and the corresponding stiffness construction. Considering the configuration shown in Fig. 6.1, where the platform has one degree of freedom along the vertical z-axis. The freedom space can be presented by twist t_1 as

$$\mathbb{S} = \left\{t_1 = \begin{bmatrix} 0 & 0 & 0 & 0 & 0 & 1 \end{bmatrix}^{\mathrm{T}} \right. \tag{6.6}$$

The corresponding constraint space can be calculated according to Eq. (5.1). The fundamental principle is to calculate the basis of the null space of the freedom space and interchange the first and second part of each basis vector [32]. For instance, an initial constraint space can be determined by calculating the orthonormal basis of the matrix in Eq. (6.6) and interchanging the first and second part of them, which can be written as

$$\mathbb{S}^c = \begin{cases} \mathbf{S}_1 = \begin{bmatrix} 1 & 0 & 0 & 0 & 0 & 0 \end{bmatrix}^{\mathrm{T}} \\ \mathbf{S}_2 = \begin{bmatrix} 0 & 1 & 0 & 0 & 0 & 0 \end{bmatrix}^{\mathrm{T}} \\ \mathbf{S}_3 = \begin{bmatrix} 0 & 0 & 0 & 1 & 0 & 0 \end{bmatrix}^{\mathrm{T}} \\ \mathbf{S}_4 = \begin{bmatrix} 0 & 0 & 0 & 0 & 1 & 0 \end{bmatrix}^{\mathrm{T}} \\ \mathbf{S}_5 = \begin{bmatrix} 0 & 0 & 0 & 0 & 0 & 1 \end{bmatrix}^{\mathrm{T}} \end{cases} \tag{6.7}$$

where $S_i (i = 1, \ldots, 5)$ is the screw associated with wrench w_i. We can easily find the obtained constraint wrenches S_1 and S_2 that resisting the rest two translations along x-axis and y-axis satisfy the criteria presented in Eq. (5.2), while S_3, S_4 and S_5 that resisting rotations about three axes do not satisfy this criteria. Additional augmentation procedures [7] are taken by adding $S_i (i = 1, 2)$ to $S_j (j = 3, 4, 5)$ in Eq. (6.7) to generate five new constraint wrenches. They are linearly independent and agree with the line constraint requirements. Also by taking account the distance r between the line constraint to the origin of coordinate frame $\{O, x, y, z\}$, one possible solution is written as

$$
\mathbb{S}^c =
\begin{cases}
S_1 = \begin{bmatrix} 1 & 0 & 0 & 0 & 0 & 0 \end{bmatrix}^T \\
S_2 = \begin{bmatrix} 0 & 1 & 0 & 0 & 0 & 0 \end{bmatrix}^T \\
S_3 = \begin{bmatrix} 0 & 1 & 0 & r & 0 & 0 \end{bmatrix}^T \\
S_4 = \begin{bmatrix} 1 & 0 & 0 & 0 & r & 0 \end{bmatrix}^T \\
S_5 = \begin{bmatrix} 1 & 0 & 0 & 0 & 0 & r \end{bmatrix}^T
\end{cases}
\tag{6.8}
$$

The schematic diagram of the configuration corresponding to Eq. (6.8) is illustrated in Fig. 6.3. There are other ways of permuting the initial five constraint wrenches and they are omitted here. Further we categorize the modified constraint wrenches in Eq. (6.8) into two groups A and B, group A contains S_1 and S_2 that resist translations and they can be obtained from initial results directly; Group B contains S_3 and S_4 and S_5 that resist rotations and they are modified basis vectors. This classification will be used in the constraint stiffness synthesis in the following sections.

Further the constraint stiffness matrix can be calculated according to Eq. (6.8). According to the physical prototype [29], the stiffness of bias springs of each line constraint is $k_l = 2500 \, \text{N/m}$ and the distance $r = 16 \, \text{mm}$. Substituting Eq. (6.8) and physical properties of each constraint into Eq. (6.3), the constraint stiffness matrix is calculated as

Fig. 6.3 Freedom and constraint space of the configuration that corresponds to Eq. (6.8)

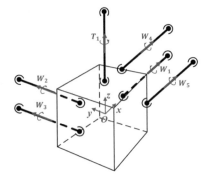

$$\mathbf{K} = \begin{bmatrix} 5000 & 0 & 0 & 0 & 40 & 0 \\ 0 & 5000 & 0 & 40 & 0 & 0 \\ 0 & 0 & 2500 & 0 & 0 & 0 \\ 0 & 40 & 0 & 0.64 & 0 & 0 \\ 40 & 0 & 0 & 0 & 0.64 & 0 \\ 0 & 0 & 0 & 0 & 0 & 0 \end{bmatrix} \tag{6.9}$$

from which we can see $trace(\mathbf{K}\Delta) = 0$. Together with the classification of constraint groups, this property is proved to be essential in the further constraint stiffness synthesis in the following sections.

6.3 Constraint Stiffness Decomposition Using the Matrix-Partition Approach

Following the construction of constraint stiffness matrix in Sect. 6.2, it is used to conduct stiffness decomposition in the following sections, which aims at finding constraint limbs with suitable layouts and elasticities that can construct the same stiffness matrix. This is also identified as a stiffness synthesis problem. As has discussed, there are mainly two types of stiffness synthesis algorithms, including the direct-recursion algorithm [13–15] and the matrix-partition algorithm [19–21, 25]. The main difference between them lies with the fact that the direct-recursion algorithm decomposes the stiffness matrix directly while matrix-partition algorithm divides the stiffness matrix into sub-matrices and decompose each of them separately.

It is revealed in Sect. 6.2.2 that two groups of constraint limbs are categorized to generate the desired motion of the functional platform. Taking the configuration discussed in Sect. 6.2.2 for example. Group A includes constraint limbs S_1 and S_2 that pass through the origin of the coordinate frame and they generate pure force to resist translational degrees of freedom; Group B contains constraint limbs S_3, S_4 and S_5 that generate wrenches to resist rotational degrees of freedom. In fact, for any given mobility, there are always existing such two groups of constraint limbs, both of which have a rank equal or less than three. This inspires us to select a suitable matrix-partition algorithm to decompose the constraint stiffness matrix into two sub-matrices according to this categorization.

6.3.1 Line-Vector Decomposition of the Constraint Stiffness Matrix

Since all constraint limbs behave as linear springs, the constraint stiffness can be decomposed using the line-vector synthesis approach proposed in [19]. The line-vector synthesis approach is developed from the line-vector eigenvalue problem discussed in [33], which is a complementary study of the free-vector eigenvalue problem [23]. In the line-vector eigenvalue problem, the concepts of co-eigenwrench and co-eigentwist are defined. A co-eigentwist causes a pure force parallel to the translation part of the co-eigentwist, and a co-eigenwrench causes a pure rotation parallel to the couple part of the co-eigenwrench. In this approach, the stiffness matrix $\mathbf{K}$ can be written in a block form as

$$\mathbf{K} = \begin{bmatrix} \mathbf{K}_{11} & \mathbf{K}_{12} \\ \mathbf{K}_{21} & \mathbf{K}_{22} \end{bmatrix} \tag{6.10}$$

where $\mathbf{K}_{11}$ is the linear part, $\mathbf{K}_{22}$ is the angular part and $\mathbf{K}_{12}$ and $\mathbf{K}_{21}$ are the coupling parts. They are all 3×3 sub-matrices. According to [33], the line-vector eigenvalue decomposition of $\mathbf{K}$ can be presented as

$$\mathbf{K} \begin{bmatrix} \boldsymbol{\delta} \\ \boldsymbol{\theta} \end{bmatrix} = \begin{bmatrix} \boldsymbol{\delta} \\ \mathbf{0} \end{bmatrix} \mathbf{k}_\delta \qquad \mathbf{K} \begin{bmatrix} \mathbf{0} \\ \mathbf{m} \end{bmatrix} = \begin{bmatrix} \mathbf{f} \\ \mathbf{m} \end{bmatrix} \mathbf{k}_m \tag{6.11}$$

$\begin{bmatrix} \boldsymbol{\delta}^\mathrm{T} & \boldsymbol{\theta}^\mathrm{T} \end{bmatrix}^\mathrm{T}$ is a 3×6 matrix whose columns $\begin{bmatrix} \boldsymbol{\delta}_i^\mathrm{T} & \boldsymbol{\theta}_i^\mathrm{T} \end{bmatrix}^\mathrm{T}$ $(i = 1, 2, 3)$ are called co-eigentwists [33]. They are written in the Plucker axis coordinates, as δ_i is the unit translation part accompanied by the rotation θ_i part. Co-eigentwists cause pure forces $\begin{bmatrix} \boldsymbol{\delta}^\mathrm{T} & \mathbf{0}^\mathrm{T} \end{bmatrix}^\mathrm{T}$, and $\mathbf{k}_\delta$ is the diagonal matrix with each diagonal element corresponding to the stiffness constant k_δ. Similarly, $\begin{bmatrix} \mathbf{f}^\mathrm{T} & \mathbf{m}^\mathrm{T} \end{bmatrix}^\mathrm{T}$ is a 3×6 matrix whose columns $\begin{bmatrix} \mathbf{f}_i^\mathrm{T} & \mathbf{m}_i^\mathrm{T} \end{bmatrix}^\mathrm{T}$ are named as co-eigenwrenches, which are written in Plucker ray coordinates. $\mathbf{m}_i$ is the unit moment part accompanied by a force part $\mathbf{f}_i$, $\mathbf{k}_m$ is the diagonal matrix with each diagonal element corresponding to the stiffness constant k_m. $\begin{bmatrix} \mathbf{f}^\mathrm{T} & \mathbf{m}^\mathrm{T} \end{bmatrix}^\mathrm{T}$ result in pure rotations $\begin{bmatrix} \mathbf{0}^\mathrm{T} & \mathbf{m}^\mathrm{T} \end{bmatrix}^\mathrm{T}$. Equation (6.11) leads to the stiffness decomposition of $\mathbf{K}$ as

$$\mathbf{K} = \begin{bmatrix} \boldsymbol{\delta}\sqrt{\mathbf{k}_\delta} \\ \mathbf{0} \end{bmatrix} \begin{bmatrix} (\boldsymbol{\delta}\sqrt{\mathbf{k}_\delta})^\mathrm{T} & \mathbf{0}^\mathrm{T} \end{bmatrix}^\mathrm{T} + \begin{bmatrix} \mathbf{f}\sqrt{\mathbf{k}_m} \\ \mathbf{m}\sqrt{\mathbf{k}_m} \end{bmatrix} \begin{bmatrix} (\mathbf{f}\sqrt{\mathbf{k}_m})^\mathrm{T} & (\mathbf{m}\sqrt{\mathbf{k}_m})^\mathrm{T} \end{bmatrix}^\mathrm{T} \tag{6.12}$$

If the first partitioned matrix is denoted as $\mathbf{K}_A$ and the second partitioned matrix is denoted as $\mathbf{K}_B$, then it is easy to identify that $\mathbf{K}_A$ satisfies Eq. (6.5) and it is a realization of three linear springs that intersect at the origin of the coordinate frame, which is equivalent to the constraint limbs that belong to group A; $\mathbf{K}_B$ is a realization of the rest linear springs that generate co-eigenwrenches in the Eq. (6.11), which are equivalent to the constraint limbs that belong to group B. Thus the decomposition of constraint stiffness matrix $\mathbf{K}$ is simplified to be the decomposition of $\mathbf{K}_A$ and $\mathbf{K}_B$.

6.3.2 Decomposition of $\mathbf{K}_A$

The explicit solution of $\mathbf{K}_A$ can be obtained by substituting Eq. (6.10) into Eq. (6.11) as

$$\begin{aligned} \mathbf{K}_{11}\boldsymbol{\delta} + \mathbf{K}_{12}\boldsymbol{\theta} &= \boldsymbol{\delta}\mathbf{k}_\delta \\ \mathbf{K}_{21}\boldsymbol{\delta} + \mathbf{K}_{22}\boldsymbol{\theta} &= \mathbf{0} \end{aligned} \tag{6.13}$$

When $\mathbf{K}_{22}$ is positive definite, Eq. (6.13) can be solved directly by substituting the second equation into the first one where $\boldsymbol{\theta} = -\mathbf{K}_{22}^{-1}\mathbf{K}_{21}\boldsymbol{\delta}$. When $rank(\mathbf{K}_{22}) < 3$, the pseudoinverse $\mathbf{K}_{22}^{+}$ is used instead of $\mathbf{K}_{22}^{-1}$ [34]. Equation (6.13) can be solved as

$$\begin{bmatrix} \boldsymbol{\delta}\sqrt{\mathbf{k}_\delta} \\ \mathbf{0} \end{bmatrix} = \begin{bmatrix} \sqrt{\mathbf{K}_{11} - \mathbf{K}_{12}\mathbf{K}_{22}^{+}\mathbf{K}_{21}}\,\mathbf{R}_1 \\ \mathbf{0} \end{bmatrix} \tag{6.14}$$

When $\mathbf{K}_{22}$ is full rank, $\boldsymbol{\delta}\sqrt{\mathbf{k}_\delta} = \sqrt{\mathbf{K}_{11} - \mathbf{K}_{12}\mathbf{K}_{22}^{-1}\mathbf{K}_{21}}\,\mathbf{R}_1$, which is also obtained in [21, 25]. $\mathbf{R}_1$ is an 3×3 orthogonal matrix, it can be chosen arbitrarily either for the purpose of geometrical properties or stiffness constants of synthesized springs. In addition, $\mathbf{K}_A$ can be calculated according to Eqs. (6.12) and (6.14), which has the form

$$\mathbf{K}_A = \begin{bmatrix} \mathbf{K}_{11} - \mathbf{K}_{12}\mathbf{K}_{22}^{+}\mathbf{K}_{21} & \mathbf{0} \\ \mathbf{0} & \mathbf{0} \end{bmatrix} \tag{6.15}$$

6.3.3 Decomposition of $\mathbf{K}_B$

Following the decomposition of the first stiffness matrix $\mathbf{K}_A$, the problem turns out to be finding the rest linear springs that construct the second stiffness matrix $\mathbf{K}_B$. $\mathbf{K}_B$ can be calculated by subtracting $\mathbf{K}_A$ from $\mathbf{K}$, which has the form

$$\mathbf{K}_B = \begin{bmatrix} \mathbf{K}_{12}\mathbf{K}_{22}^{+}\mathbf{K}_{21} & \mathbf{K}_{12} \\ \mathbf{K}_{21} & \mathbf{K}_{22} \end{bmatrix} \tag{6.16}$$

Substituting Eq. (6.10) into Eq. (6.11), we can obtain

$$\begin{bmatrix} \mathbf{f}\sqrt{\mathbf{k}_m} \\ \mathbf{m}\sqrt{\mathbf{k}_m} \end{bmatrix} = \begin{bmatrix} \mathbf{K}_{12}\sqrt{\mathbf{K}_{22}^{+}}\mathbf{R}_2 \\ \sqrt{\mathbf{K}_{22}}\mathbf{R}_2 \end{bmatrix} \tag{6.17}$$

which suggest we should determine the suitable $\mathbf{R}_2$ that satisfying the isotropic condition

$$\mathbf{R}_2^T (\sqrt{\mathbf{K}_{22}}\mathbf{K}_{12}\sqrt{\mathbf{K}_{22}^+} + \sqrt{\mathbf{K}_{22}^+}\mathbf{K}_{21}\sqrt{\mathbf{K}_{22}})\mathbf{R}_2 = \begin{bmatrix} 0 & \bullet & \bullet \\ \bullet & 0 & \bullet \\ \bullet & \bullet & 0 \end{bmatrix} \qquad (6.18)$$

Equation (6.18) can be solved using the direct-recursion methods [14, 15]. Since $rank(\mathbf{K}_B) \leq 3$, closed form solutions are feasible and they can be referred alternatively to [21, 35]. If the matrix in Eq. (6.18) is denoted as $\mathbf{H}$, then $\mathbf{H}$ has exactly one or two isotropic vectors [35] when $rank(\mathbf{K}_b) = 1$ or $rank(\mathbf{K}_b) = 2(\mathbf{H} \neq \mathbf{0})$. When $rank(\mathbf{K}_b) = 2(\mathbf{H} = \mathbf{0})$ or $rank(\mathbf{K}_b) = 3$, $\mathbf{H}$ has infinite isotropic vectors, which enables us to select suitable $\mathbf{R}_2$ from the closed-form solution and design preferred linear springs.

In conclusion, the line-vector decomposition method can divide the stiffness matrix $\mathbf{K}$ into two sub-matrices $\mathbf{K}_A$ and $\mathbf{K}_B$. $\mathbf{K}_A$ is the integration of linear springs that pass through the origin and is equivalent to the integration of constraint limbs that resist translational degrees of freedom; The second one $\mathbf{K}_B$ is the integration of the rest linear springs which are equivalent to the constraint limbs that resist rotational degrees of freedom. Both $\mathbf{K}_A$ and $\mathbf{K}_B$ can be decomposed separately by selecting suitable $\mathbf{R}_1$ and $\mathbf{R}_2$ discussed in Sects. 6.3.2 and 6.3.3.

6.4   Implementation of Constraint-Stiffness Decomposition Algorithms

In this section, both matrix-partition and direct-recursion algorithms are used to synthesize the constraint stiffness matrix of SPS parallel mechanism. The matrix-partition algorithm is developed from the line-vector synthesis approach discussed in Sect. 6.3. Since the constraint limbs have the same stiffness coefficients in the SPS-parallel platform, to effectively compare the decomposition result and initial layout of them, $\mathbf{R}_1$ and $\mathbf{R}_2$ are selected to equalize the stiffness constant of decomposed constraint limbs in the developed matrix-partition approach. The direct-recursion algorithm developed in [14] is used to decompose the constraint stiffness matrix as a comparison result. Moreover, the rank of the constraint stiffness matrix changes according to the platform's mobility, thus various types of them with rank equal or less than six are examined.

6.4.1   Rank-Six Constraint Stiffness Matrix Decomposition

The constraint stiffness matrix with functional platform fully constrained is studied first. The process of calculating the layouts of constraint limbs is provided in Sect. 6.2. One possible configuration is written as

Fig. 6.4 Original layout of
rank-6 constraints

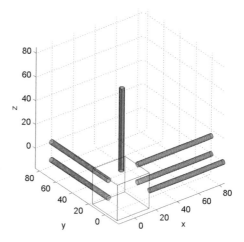

$$
\mathbb{S}^c =
\begin{cases}
S_1 = \begin{bmatrix} 1 & 0 & 0 & 0 & 0 & 0 \end{bmatrix}^T \\
S_2 = \begin{bmatrix} 1 & 0 & 0 & 0 & 0 & r \end{bmatrix}^T \\
S_3 = \begin{bmatrix} 1 & 0 & 0 & 0 & r & 0 \end{bmatrix}^T \\
S_4 = \begin{bmatrix} 0 & 1 & 0 & 0 & 0 & 0 \end{bmatrix}^T \\
S_5 = \begin{bmatrix} 0 & 1 & 0 & r & 0 & 0 \end{bmatrix}^T \\
S_6 = \begin{bmatrix} 0 & 0 & 1 & 0 & 0 & 0 \end{bmatrix}^T
\end{cases}
\tag{6.19}
$$

which is a "3-2-1" type configuration [36]. The corresponding layout of constraints
is shown in Fig. 6.4. According to the physical prototype, the stiffness of bias springs
of each line constraint is $k_l = 2500\,\text{N/m}$. For constraints S_2, S_3 and S_5, the vertical
distance between their constraint axes and origin is 16mm. Substituting Eq. (6.19)
and these physical properties into Eq. (6.3), the corresponding constraint stiffness
matrix is calculated as

$$
K =
\begin{bmatrix}
7500 & 0 & 0 & 0 & 40 & 40 \\
0 & 5000 & 0 & 40 & 0 & 0 \\
0 & 0 & 2500 & 0 & 0 & 0 \\
0 & 40 & 0 & 0.64 & 0 & 0 \\
40 & 0 & 0 & 0 & 0.64 & 0 \\
40 & 0 & 0 & 0 & 0 & 0.64
\end{bmatrix}
\tag{6.20}
$$

It is easy to identify that $trace(K\Delta) = 0$ and $rank(K) = 6$. Subsequently both
direct-recursion method and matrix-partition method are applied to synthesize the
constraint stiffness matrix in Eq. (6.20). The direct-recursion method is used without
any input, while the matrix-partition method is applied with input to equalize the

stiffness constants of each decomposed spring. For instance, the divided stiffness matrix $\mathbf{K}_A$ can be calculated according to Eq. (6.15) as

$$
\mathbf{K}_A = \begin{bmatrix}
2500 & 0 & 0 & 0 & 0 & 0 \\
0 & 2500 & 0 & 0 & 0 & 0 \\
0 & 0 & 2500 & 0 & 0 & 0 \\
0 & 0 & 0 & 0 & 0 & 0 \\
0 & 0 & 0 & 0 & 0 & 0 \\
0 & 0 & 0 & 0 & 0 & 0
\end{bmatrix}
\tag{6.21}
$$

which is the combination of constraint wrenches S_1, S_4 and S_6. To ensure they have the same stiffness constants, it is easy to decide that $\mathbf{R}_1$ in Eq. (6.14) is equal to $\mathbf{I}_3$ in this case. $\mathbf{K}_B$ can be calculated according to Eq. (6.16) as

$$
\mathbf{K}_B = \begin{bmatrix}
5000 & 0 & 0 & 0 & 40 & 40 \\
0 & 2500 & 0 & 40 & 0 & 0 \\
0 & 0 & 0 & 0 & 0 & 0 \\
0 & 40 & 0 & 0.64 & 0 & 0 \\
40 & 0 & 0 & 0 & 0.64 & 0 \\
40 & 0 & 0 & 0 & 0 & 0.64
\end{bmatrix}
\tag{6.22}
$$

which is the combination of constraint wrenches S_2, S_3 and S_5. They can be further calculated according to Eq. (6.17), which have the forms

$$
\begin{bmatrix} \mathbf{f}\sqrt{k_m} \\ \mathbf{m}\sqrt{k_m} \end{bmatrix} = \begin{bmatrix}
50 & 0 & 0 & 0 & 0 & 0.8 \\
50 & 0 & 0 & 0 & 0.8 & 0 \\
0 & 50 & 0 & 0.8 & 0 & 0
\end{bmatrix}^{\mathrm{T}}
\tag{6.23}
$$

Also $\mathbf{R}_2$ in Eq. (6.18) is an identity matrix $\mathbf{I}_3$ so that the stiffness constants of synthesized results are the same. From Eq. (6.23) we can see the synthesized configuration of S_2, S_3 and S_5 are exactly the same as those used in Eq. (6.19). Also the original constraint configuration of the parallel platform and synthesized results are shown in Fig. 6.5. The decomposed constraint limbs from matrix-partition method are shown in Fig. 6.5a and those generated by direct-recursion method are shown in Fig. 6.5b. Both methods generate similar geometrical patterns of decomposed constraint limbs, as three constraint vectors are mutually parallel to the x axis, they constraint degrees of freedom including translation along x-axis and rotations about x and z axes. The other three constraint wrenches are lying in the $y - z$ plane, they do not intersect at a common point and constraint the motions including translation along y and z axes and rotation about y axis. These two sub-constraint space are ranking three and they together construct a rank-six constraint space.

However, the stiffness constants and positions of synthesized constraints with two methods are different, which are listed in Table 6.1. The matrix-partition method generates six constraint limbs with stiffness constant 2500 N/m, the positions and orientations of which are exactly the same as those provided in Eq. (6.19). These lie

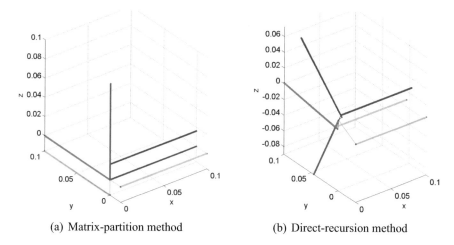

(a) Matrix-partition method   (b) Direct-recursion method

Fig. 6.5 Comparisons of synthesized results when $rank(\mathbf{K}) = 6$

Table 6.1 Comparisons of synthesized springs when $rank(\mathbf{K}) = 6$

Matrix-partition approach				Direct-recursion approach		
No.	Stiffness (N/m)	Orientation (rad)	Position (mm)	Stiffness (N/m)	Orientation (rad)	Position (mm)
1	2500	$[1, 0, 0]$	$[0, 0, 0]$	6481	$[1, 0, 0]$	$[0, -4.2, 8.3]$
2	2500	$[1, 0, 0]$	$[0, -16, 0]$	500	$[1, 0, 0]$	$[0, -32, -16]$
3	2500	$[1, 0, 0]$	$[0, 0, 16]$	519	$[1, 0, 0]$	$[0, 5.6, -11.2]$
4	2500	$[0, 1, 0]$	$[0, 0, -16]$	3991	$[0, 0.99, 0.13]$	$[0, 1.5, -11.7]$
5	2500	$[0, 1, 0]$	$[0, 0, 0]$	2500	$[0, 0.45, 0.89]$	$[0, 0, 0]$
6	2500	$[0, 0, 1]$	$[0, 0, 0]$	1009	$[0, 0.75, 0.66]$	$[0, -5.9, 6.8]$

with in the fact that the constraint wrenches listed in Eq. (6.19) have one to one correspondence with the decomposed linear springs using this matrix-partition approach.
Compared with the matrix-partition approach, the direct-recursion method generates six constraint limbs with similar geometrical patterns, but the stiffness constant
varies from 500 to 6481 N/m. This will result in different geometrical dimensions of
each bias spring and increase difficulty in assembling them into the constraint limbs.
Meanwhile, the positions and orientations of them are also varying from 4.2 to
32 mm, which may cause potential spacing difficulties. However, the sum of synthesized spring stiffness constants with these two methods are both equal to 15,000 N/m,
which is equal to the $trace(\mathbf{K}_{11})$.

6.4.2   *Rank-Five Constraint Stiffness Matrix Decomposition*

Following the study of the fully constrained platform, the constraint-stiffness matrix with the platform having some degrees of freedom is studied in this section. Without loss of generality, the platform is assumed to be able to rotate freely about z axis. The corresponding constraint space is written as

$$\mathbb{S}^c = \begin{cases} S_1 = \begin{bmatrix} 1 & 0 & 0 & 0 & 0 & 0 \end{bmatrix}^T \\ S_2 = \begin{bmatrix} 1 & 0 & 0 & 0 & r & 0 \end{bmatrix}^T \\ S_3 = \begin{bmatrix} 0 & 1 & 0 & 0 & 0 & 0 \end{bmatrix}^T \\ S_4 = \begin{bmatrix} 0 & 1 & 0 & r & 0 & 0 \end{bmatrix}^T \\ S_5 = \begin{bmatrix} 0 & 0 & 1 & 0 & 0 & 0 \end{bmatrix}^T \end{cases} \tag{6.24}$$

The corresponding layout of constraints is shown in Fig. 6.6. The stiffness constant of each constraint and the geometrical properties are the same as those used in Sect. 6.4.1. Substituting Eq. (6.24) and physical properties of each constraint into Eq. (6.3), the stiffness matrix of constraints is calculated as

$$\mathbf{K} = \begin{bmatrix} 5000 & 0 & 0 & 0 & 40 & 0 \\ 0 & 5000 & 0 & 40 & 0 & 0 \\ 0 & 0 & 2500 & 0 & 0 & 0 \\ 0 & 40 & 0 & 0.64 & 0 & 0 \\ 40 & 0 & 0 & 0 & 0.64 & 0 \\ 0 & 0 & 0 & 0 & 0 & 0 \end{bmatrix} \tag{6.25}$$

Fig. 6.6 Original layout of rank-5 constraints

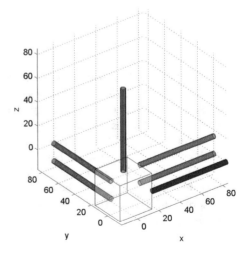

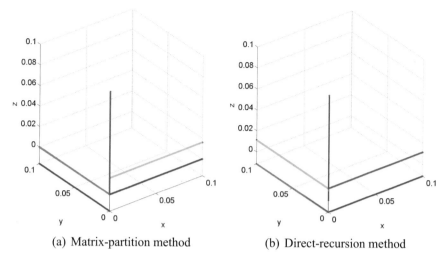

(a) Matrix-partition method   (b) Direct-recursion method

Fig. 6.7 Comparisons of synthesized results when $rank(\mathbf{K}) = 5$

from which we can see $trace(\mathbf{K}\Delta) = 0$ and $rank(\mathbf{K}) = 5$, and $\mathbf{K}$ is no longer a positive-definite matrix. Applying both the direct-recursion and matrix-partition methods, we can obtain two synthesized results, and they are presented in Fig. 6.7 together with the initial configuration of constraint limbs. Figure 6.7a demonstrates the geometrical pattern of synthesized constraint limbs using matrix-partition approach, which utilizes the design criterion that requires the stiffness constant to be equal. Again the synthesized results are the same as the geometrical pattern provided by Eq. (6.24). Figure 6.7b demonstrates the configurations of decomposed constraints using direct-recursion approach. A similar geometrical pattern can be identified, as two constraint vectors lie in the $x - z$ plane, and three lie in the $y - z$ plane that are not intersecting at a common point. The sub-constraint space occupying $x - z$ plane is ranking two, and the one occupying $y - z$ plane is ranking three, they together construct a rank-five constraint space.

The stiffness constant and geometrical properties of synthesized results are listed in Table 6.2. The matrix-partition method with input requiring the equal stiffness constant generate five constraints with the stiffness constant 2500 N/m. The stiffness constant generated with the direct-recursion method varies from 732 to 4268 N/m. There is one identical decomposed spring with two methods, which is the one parallel to z axis and passing via the origin. Also, the sub-constraint space formed by springs lying in either $x - z$ plane or $y - z$ plane are also identical and can replace each other with the two different approaches. In addition, the sum of synthesized spring stiffness constant with these two methods are 12,500 N/m, which also equal to $trace(\mathbf{K}_{11})$. Further, a more complete synthesis comparison of selected constraint spaces is provided in Appendix B that demonstrates the effectiveness of the proposed matrix-partition algorithm.

Table 6.2 Comparisons of synthesized springs when $rank(\mathbf{K}) = 5$

	Matrix-partition approach			Direct-recursion approach		
No.	Stiffness (N/m)	Orientation (rad)	Position (mm)	Stiffness (N/m)	Orientation (rad)	Position (mm)
1	2500	$[1, 0, 0]$	$[0, 0, 0]$	4268	$[1, 0, 0]$	$[0, 0, 11.3]$
2	2500	$[1, 0, 0]$	$[0, 16, 0]$	732	$[1, 0, 0]$	$[0, 0, -11.3]$
3	2500	$[0, 1, 0]$	$[0, 0, -16]$	732	$[0, 1, 0]$	$[0, 0, 11.3]$
4	2500	$[0, 1, 0]$	$[0, 0, 0]$	4268	$[0, 1, 0]$	$[0, 0, -11.3]$
5	2500	$[0, 0, 1]$	$[0, 0, 0]$	2500	$[0, 0, 1]$	$[0, 0, 0]$

6.5   Conclusions

In conclusion, this chapter systematically investigates both stiffness analysis and synthesis problems using the constraint stiffness matrix of a SPS-orthogonal compliant parallel mechanism. In the stiffness analysis, the layouts of the constraint limbs are determined according to the reciprocal relationships between motions and constraint wrenches. The constraint stiffness matrix is then developed using their spatial geometrical and material properties. Some intrinsic properties are identified, including the isotropic condition as well as properties of symmetric matrices, which are used in the constraint stiffness synthesis process.

To solve the stiffness synthesis problem, the current synthesis approaches are compared and categorized into two types, including the direct-recursion type and matrix-partition type. Direct-recursion approaches are purely algebra-based algorithms that generate synthesized springs automatically from the original stiffness matrix; the matrix-partition methods divide the stiffness matrix at the beginning and decompose each of the sub-matrices using similar recursive algorithms. Compared to the direct-recursion approaches, matrix-partition approaches consider the properties of synthesized springs and utilize them as selection criteria, which provide some design flexibilities. A further exploration of the matrix-partition approach is conducted. Since one constraint limb has a passive structure that compliantly resists the motion along the axis and behaves as a linear spring, an one-to-one correspondence relationship between them and synthesized results using the line-vector eigenvalue decomposition are established.

Subsequently both the line-vector based synthesis algorithm as well as a direct-recursion algorithm are used to decompose the constraint stiffness matrix. Various types of constraint layouts are selected, which result in stiffness matrices with different ranks. The synthesis results suggest both the selected direct-recursion algorithm and matrix-partition algorithm can decompose the constraint stiffness matrix with minimum numbers of decomposed linear springs, and the sum of their stiffness coefficients are also equal. Particularly, by setting the design criterion to equalize the stiffness constants, the modified matrix-partition approach can synthesize the constraint stiffness with exactly the same geometrical configurations of constraint limbs used in the compliant parallel mechanisms.

The main contributions of this chapter include: (1) uniquely uses a physical prototype of one compliant parallel mechanism to demonstrate both the stiffness analysis and synthesis procedures; (2) compares the existing stiffness synthesis approaches and categorize them as direct-recursion algorithms and matrix-partition algorithms; (3) Compares and contrasts the effectiveness of both types of algorithms using the constraint stiffness matrix of the parallel platform, and particularly establishes a one-to-one correspondence between the synthesized results and the initial configurations of constraint limbs used to construct the stiffness matrix. This presented work has potential applications in the structural construction and optimization of force sensors, as well as robotic hand grasping synthesis using grasping stiffness properties.

References

1. Bhattacharya, S., Hatwal, H., Ghosh, A.: On the optimum design of stewart platform type parallel manipulators. Robotica **13**(02), 133–140 (1995)
2. Dwarakanath, T., Dasgupta, B., Mruthyunjaya, T.: Design and development of a stewart platform based force-torque sensor. Mechatronics **11**(7), 793–809 (2001)
3. Kong, X., Gosselin, C.: Type Synthesis of Parallel Mechanisms, vol. 33. Springer (2007)
4. Blanding, D.L.: Exact constraint: machine design using kinematic processing. American Society of Mechanical Engineers (1999)
5. Dai, J.S., Kerr, D.: A six-component contact force measurement device based on the Stewart platform. Proc. Inst. Mech. Eng. Part C: J. Mech. Eng. Sci. **214**(5), 687–697 (2000)
6. Hopkins, J.B., Panas, R.M.: Design of flexure-based precision transmission mechanisms using screw theory. Precis. Eng. **37**(2), 299–307 (2013)
7. Yu, J.J., Li, S.Z., Qiu, C.: An analytical approach for synthesizing line actuation spaces of parallel flexure mechanisms. J. Mech. Des. **135**(12), 124501–124501 (2013)
8. Dai, J.S., Rees Jones, J.: Interrelationship between screw systems and corresponding reciprocal systems and applications. Mech. Mach. Theory **36**(5), 633–651 (2001)
9. Dai, J.S., Jones, J.R.: Null–space construction using cofactors from a screw–algebra context. Proc. Royal Soc. Lond. Ser. A: Math. Phys. Eng. Sci. **458**(2024), 1845–1866 (2002)
10. Ball, R.S.: A Treatise on the Theory of Screws. Cambridge University Press (1900)
11. Dimentberg, F.M.: The screw calculus and its applications in mechanics. Technical report, DTIC Document (1968)
12. Loncaric, J.: Normal forms of stiffness and compliance matrices. IEEE J. Robot. Autom. **3**(6), 567–572 (1987)
13. Huang, S., Schimmels, J.M.: The bounds and realization of spatial stiffnesses achieved with simple springs connected in parallel. IEEE Trans. Robot. Autom. **14**(3), 466–475 (1998)
14. Ciblak, N., Lipkin, H.: Synthesis of cartesian stiffness for robotic applications. In: Proceedings of the IEEE International Conference on Robotics and Automation, vol. 3, pp. 2147–2152. IEEE (1999)
15. Roberts, R.G.: Minimal realization of a spatial stiffness matrix with simple springs connected in parallel. IEEE Trans. Robot. Autom. **15**(5), 953–958 (1999)
16. Dai, J.S., Xilun, D.: Compliance analysis of a three-legged rigidly-connected platform device. J. Mech. Des. **128**(4), 755–764 (2006)
17. Ding, X., Dai, J.S.: Characteristic equation-based dynamics analysis of vibratory bowl feeders with three spatial compliant legs. IEEE Trans. Autom. Sci. Eng. **5**(1), 164–175 (2008)
18. Dai, J.S.: Geometrical Foundations and Screw Algebra for Mechanisms and Robotics. Higher Education Press, Beijing (2014). ISBN 9787040334838. (translated from Dai, J.S.: Screw Algebra and Kinematic Approaches for Mechanisms and Robotics. Springer, London (2016))

19. Ciblak, N., Lipkin, H.: Application of Stiffness Decompositions to Synthesis by Springs. ASME (1998)
20. Choi, K., Jiang, S., Li, Z.: Spatial stiffness realization with parallel springs using geometric parameters. IEEE Trans. Robot. Autom. **18**(3), 274–284 (2002)
21. Roberts, R.G., Shirey, T.: Algorithms for passive compliance mechanism design. In: Proceedings of the 35th Southeastern Symposium on System Theory, 2003, pp. 347–351. IEEE (2003)
22. Chen, G., Wang, H., Lin, Z., Lai, X.: The principal axes decomposition of spatial stiffness matrices. IEEE Trans. Robot. **31**(1), 191–207 (2015)
23. Lipkin, H., Patterson, T.: Geometrical properties of modelled robot elasticity: Part I-decomposition. In: 1992 ASME Design Technical Conference. Scottsdale, DE, vol. 45, pp. 179–185 (1992)
24. Huang, S., Schimmels, J.M.: The eigenscrew decomposition of spatial stiffness matrices. IEEE Trans. Robot. Autom. **16**(2), 146–156 (2000)
25. Huang, S., Schimmels, J.M.: Minimal realizations of spatial stiffnesses with parallel or serial mechanisms having concurrent axes. J. Robotic Syst. **18**(3), 135–146 (2001)
26. Roberts, R.G.: Minimal realization of an arbitrary spatial stiffness matrix with a parallel connection of simple and complex springs. IEEE Trans. Robot. Autom. **16**(5), 603–608 (2000)
27. Griffis, M., Duffy, J.: Global stiffness modeling of a class of simple compliant couplings. Mech. Mach. Theory **28**(2), 207–224 (1993)
28. Ciblak, N., Lipkin, H.: Asymmetric cartesian stiffness for the modeling of compliant robotic systems. In: Proceedings of the 23rd Biennial ASME Mechanisms Conference, Minneapolis, MN (1994)
29. Qiu, C., Zhang, K., Dai, J.S.: Constraint-based design and analysis of a compliant parallel mechanism using SMA-spring actuators. In: ASME 2014 International Design Engineering Technical Conferences and Computers and Information in Engineering Conference, pp. V05AT08A035–V05AT08A035. American Society of Mechanical Engineers (2014)
30. Gosselin, C.: Stiffness mapping for parallel manipulators. IEEE Trans. Robot. Autom. **6**(3), 377–382 (1990)
31. Li, Y., Xu, Q.: Stiffness analysis for a 3-PUU parallel kinematic machine. Mech. Mach. Theory **43**(2), 186–200 (2008)
32. Dai, J.S., Jones, J.R.: A linear algebraic procedure in obtaining reciprocal screw systems. J. Robotic Syst. **20**(7), 401–412 (2003)
33. Ciblak, N.: Analysis of cartesian stiffness and compliance with applications (1998)
34. Roberts, R.G.: Note on the normal form of a spatial stiffness matrix. IEEE Trans. Robot. Autom. **17**(6), 968–972 (2001)
35. Ciblak, N., Lipkin, H.: Orthonormal isotropic vector bases. In: Proceedings of DETC, vol. 98. Citeseer (1998)
36. Geng, Z.J., Haynes, L.S.: A 3-2-1 kinematic configuration of a Stewart platform and its application to six degree of freedom pose measurements. Robot. Comput. Integr. Manuf. **11**(1), 23–34 (1994)

Chapter 7
Compliance Parameterization and Optimization of Compliant Parallel Mechanisms

7.1 Introduction

An ortho-planar spring is a kind of compliant mechanisms [1] that utilizes the out-of-plane deformation of its flexible limbs. The ortho-planar spring has many advantages, including the capability of being fabricated from a single piece of material that reduces manufacturing costs with its compact form, enabling it to be used in a highly confined space. Parise, Howell and Magleby et al. first introduced the concept of ortho-planar-spring [2, 3] in 2001 and applied it to designing drive clutches [4] and novel lamina emergent mechanisms [5]. Nguyen and Smal et al. [6, 7] used similar planar springs in micro-valve control. Recently Ataollahi et al. [8] adopted a planar-spring structure to designing an optical-fibre based force sensor. In these applications, the out-of-plane linear compliance of ortho-planar springs is utilized and modelled by deformation of cantilever beams under a free-end load, using either the Euler-Bernoulli beam theory or the Pseudo-rigid-body model [1].

Further observation reveals that an ortho-planar spring consists of a moving platform and a base with both being connected by flexure elements, which is similar to the configuration of a parallel mechanism. As a result, we can take an ortho-planar spring as an equivalent to a parallel mechanism and develop its six-dimensional compliance matrix accordingly. The developed compliance matrix enables us to evaluate both linear and rotational compliances of an ortho-planar spring, which are essential in the practical application where a structure usually experiences more than one type of load. This investigation of the rotational compliance may also lead to innovative designs such as elastic joints and continuum structures where the rotational compliance plays an important role.

Investigation of the compliance and stiffness of parallel mechanisms can be dated back to the study of elastically suspended robotic systems [9–11] in the 1960s, and a subsequent wire-suspended platform [12] and flexure-based compliant mechanisms [13, 14]. In these studies, the platform's compliance/stiffness matrix was

© The Editor(s) (if applicable) and The Author(s), under exclusive license to Springer Nature Switzerland AG 2021
C. Qiu and J. S. Dai, *Analysis and Synthesis of Compliant Parallel Mechanisms—Screw Theory Approach*, Springer Tracts in Advanced Robotics 139, https://doi.org/10.1007/978-3-030-48313-5_7

established based on the integration of elastic limbs and used for further evaluation in stiffness mapping [15] and dimensionless study [16]. The elastic limbs include linear and torsional springs, slender beams and blades. By using screw theory, these elastic limbs are modelled in forms of screws and their integrated compliance performance are addressed in the framework of screw theory. For example, a compliance device [13] was built based on the parallel integration of slender beams, able to utilize the remote-center compliance for assembly. A vibratory bowl feeder was modelled as a parallel mechanism with leaf-spring compliance legs [17], leading to the evaluation of its compliance characteristics [17, 18]. This suggests the compliance behaviour of ortho-planar springs can be evaluated in a similar manner using the mechanism-equivalent approach [17]. Other methods of modelling compliant mechanisms include the finite-element-method based approaches [19, 20] which were used when shapes of supporting limbs are complex but demanded high computational cost. In terms of simple shape of limb elements such as slender beams and blades, analytical models are computationally efficient and can reveal the intrinsic characteristics of a compliant mechanism.

As a result, this paper investigates the six-dimensional compliance of an ortho-planar spring by treating it as a parallel mechanism and develops its compliance matrix, leading to novel findings and applications of its spatial compliance properties. In the frame work of screw theory, the compliance matrix of an ortho-planar spring is first developed utilizing the hybrid combination of elastic beam elements and is symbolically formalized. Subsequently, the diagonal compliance elements are utilized to analyze compliance variations of an ortho-planar spring and are further validated with both FEM simulation and experimental tests. The FEM simulation evaluates a total number of 30 types of planar-spring models using a multi-body based modelling approach. In the experiment test, the typical side-type and radial-type planar springs are examined and further utilized to develop a planar-spring module based continuum manipulator.

7.2 Compliance Matrix of an Ortho-Planar Spring Based on the Equivalent Parallel Mechanism

An ortho-planar spring consists of one functional platform supported by three limbs in a symmetric arrangement, each limb is a serial chain of two elastic segments in the form of slender beam elements. The whole ortho-planar spring can be considered as a parallel mechanism, and its compliance matrix is built by developing the compliance matrix of each limb first.

7.2.1 The Compliance Matrix of a Limb

A limb of the ortho-planar spring is illustrated in Fig. 7.1, and regarded as a serial arrangement of two beams i $(i = 1, 2)$. Assuming a small deflection occurs at one beam, the classic Euler-Bernoulli beam theory is used to derive the compliance matrix. If the local coordinate frame $\{C_i, x_i, y_i, z_i\}$ coincident with the midspan of the beam is used to describe the compliance matrix, beams 1 and 2 have the same compliance matrix in a diagonal form as

$$\begin{aligned} \mathbf{C} &= diag\left[C_1\ C_2\ C_3\ C_4\ C_5\ C_6 \right] \\ &= diag\left[\frac{L}{EA}\ \frac{L^3}{12EI_z}\ \frac{L^3}{12EI_y}\ \frac{L}{GI_x}\ \frac{L}{EI_y}\ \frac{L}{EI_z} \right] \end{aligned} \tag{7.1}$$

where beam i is assumed to have a constant rectangular cross-section bh and a length of L. The shear effect and inertia of rotation of the beam section are ignored. Moments of inertia I_y and I_z are given as $I_y = \frac{1}{12}hb^3$ and $I_z = \frac{1}{12}bh^3$, and torsion constant I_x for the rectangular section is given by [21] as $I_x = bh^3(\frac{1}{3} - 0.21\frac{h}{b}(1 - \frac{h^4}{12b^4}))$ when $\frac{h}{b} \leq 1$. Two sections of slender beams are rigidly fixed to a connector. The effective length of the connector is d and is generally much smaller than beam length L, that an ortho-planar spring can have a relatively compact form [2] and the compliance of the connector is ignored. A global coordinate frame $\{O_1, x, y, z\}$ is located at the end of beam 1. When external wrench $\mathbf{W}$ is applied at the free end of the limb, a deflection twist $\mathbf{T}$ is generated depending on the integrated compliance $\mathbf{C}_e$ of the limb. Based on screw theory, the deflection twist is presented with Plücker axis coordinates and the external wrench is described with Plücker ray coordinates [22, 23]. The relationship between $\mathbf{C}_e$ and $\mathbf{C}$ can be written as

$$\mathbf{C}_e = \sum_{i=1}^{m} \mathbf{Ad}_{ie}^{\mathrm{T}} \mathbf{C} \mathbf{Ad}_{ie} \tag{7.2}$$

Fig. 7.1 A limb of the ortho-planar spring

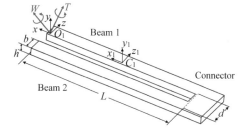

Where $\mathbf{Ad}_{ie}(i = 1, 2)$ is the adjoint transformation matrix [24] between local coordinate frame $\{C_i, x_i, y_i, z_i\}$ and global coordinate frame $\{O_1, x, y, z\}$, it has the form

$$\mathbf{Ad}_{ie} = \begin{bmatrix} \mathbf{R}_{ie} & \mathbf{0} \\ \mathbf{P}_{ie}\mathbf{R}_{ie} & \mathbf{R}_{ie} \end{bmatrix} \tag{7.3}$$

$\mathbf{R}_{ie}$ is the coordinate rotation matrix, and $\mathbf{P}_{ie}$ is the skew-symmetric matrix of the coordinate translation vector $\mathbf{p}_{ie}$. For beam $i (i = 1, 2)$, $\mathbf{R}_{ie} = \mathbf{R}_y(\frac{\pi}{2}[1 + (-1)^i])$ and $\mathbf{p}_{ie} = \left[(-1)^{i-1}\frac{L}{2} \quad 0 \quad (-1)^{i-1}(i-1)d\right]^T$. Then $\mathbf{Ad}_{ie}$ is calculated according to Eq. (7.3) and used to develop $\mathbf{C}_e$ according to Eq. (7.2). Further the stiffness matrix $\mathbf{K}_e$ of each limb can be developed and $\mathbf{K}_e = \mathbf{C}_e^{-1}$.

7.2.2 Diagonal Compliance Matrix of an Ortho-Planar Spring

Equivalent to a parallel mechanism, compliance matrix $\mathbf{C}_b$ of the ortho-planar spring can be derived based on the parallel integration of three limbs. Figure 7.2 shows the diagram of one ortho-planar spring, where each limb has one end fixed on the base and the other end connected to the functional platform B. A fixed global coordinate frame $\{B, x, y, z\}$ is defined at the center of the platform, and $\{O_i, x, y, z\}$ is the local coordinate frame attached at the end of limb i. When external wrench W is applied at the platform, deflection twist T is generated according to the compliance matrix $\mathbf{C}_b$, which is the inverse of stiffness matrix $\mathbf{K}_b$. $\mathbf{K}_b$ is written as

$$\mathbf{K}_b = \sum_{i=1}^{3} \mathbf{Ad}_{ib}^{-1}\mathbf{K}_e\mathbf{Ad}_{ib}^{-T} \tag{7.4}$$

where $\mathbf{K}_e$ is the stiffness matrix of each limb. $\mathbf{Ad}_{ib}$ is the adjoint transformation matrix between local coordinate frame $\{O_i, x, y, z\}$ and global coordinate frame $\{B, x, y, z\}$ and can be calculated the same as $\mathbf{Ad}_{ie}$. $\mathbf{R}_{ib}$ is the rotation matrix from $\{O_i, x, y, z\}$ to $\{B, x, y, z\}$, and $\mathbf{P}_{ib}$ is the anti-symmetric matrix representation for coordinate translation vector $\mathbf{p}_{ib}$. According to geometric parameters shown in Fig. 7.2, $\mathbf{R}_{ib} = \mathbf{R}_y(\frac{2\pi}{3}(i-1))$ and $\mathbf{p}_{ib} = \begin{bmatrix} m & 0 & n \end{bmatrix}^T$. Substituting stiffness matrix $\mathbf{K}_e$ and expressions of $\mathbf{R}_{ib}$ and $\mathbf{p}_{ib}$ into Eq. (7.4), we can obtain the symbolic formula of $\mathbf{K}_b$ and compliance matrix $\mathbf{C}_b$ according to $\mathbf{C}_b = \mathbf{K}_b^{-1}$, which results in a 6×6 diagonal matrix as

$$\mathbf{C}_b = diag \begin{bmatrix} C_{b1} & C_{b2} & C_{b1} & C_{b3} & C_{b4} & C_{b3} \end{bmatrix} \tag{7.5}$$

from which we can see the platform has the same linear compliance C_{b1} along both x and z axes of coordinate frame $\{B, x, y, z\}$ due to symmetric arrangement of three limbs, this is named the linear in-plane compliance since it is within the ortho-planar

Fig. 7.2 Isometric view of
the ortho-planar spring
model with three limbs

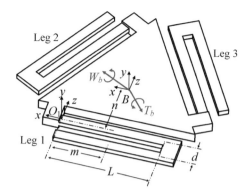

spring plane; C_{b2} is the linear out-of-plane compliance in the y direction. In terms
of rotational compliance elements, an ortho-planar spring has the same rotational
bending compliance C_{b3} about x and z axes; C_{b4} represents rotational torsional
compliance about the y axis. In order to develop the symbolic formulations of C_{bi},
nondimensional parameters are introduced as C_{bi} in a more compact form as

$$C_{b1} = C_3 C_{b1}^*, \ C_{b2} = C_2 C_{b2}^*$$
$$C_{b3} = C_6 C_{b3}^*, \ C_{b4} = C_5 C_{b4}^* \tag{7.6}$$

where C_i are the diagonal elements of single beam's compliance matrix **C** in Eq. (7.1),
The second components are symbolized as C_{bi}^* and can be written as

$$C_{b1}^* = \left\{ \frac{4\left(\frac{L}{b}\right)^2 \left(\frac{d}{L}\right)^2 + \frac{4}{3}}{\left(\frac{L}{b}\right)^2 \left[3\left(\frac{d}{L}\right)^2 + 1\right] + 1} \right\}$$

$$C_{b2}^* = \left\{ 2\left(\frac{EI_z}{GI_x}\right)\left(\frac{d}{L}\right)^2 + \frac{2}{3} \right\}$$

$$C_{b3}^* = \left\{ \frac{4\left(\frac{EI_z}{GI_x}\right)\left(\frac{d}{L}\right)^2 + \frac{4}{3}}{\left[3\left(\frac{EI_z}{GI_x}\right) + 6\right]\left(\frac{d}{L}\right)^2 + 12\left(\frac{n}{L}\right)\left(\frac{d}{L}\right) + \left[12\left(\frac{m}{L}\right)^2 + 12\left(\frac{n}{L}\right)^2 + 12\left(\frac{m}{L}\right) + 4 + \frac{GI_x}{EI_z}\right]} \right\}$$

$$C_{b4}^* = \left\{ \frac{2\left(\frac{L}{b}\right)^2 \left(\frac{d}{L}\right)^2 + \frac{2}{3}}{3\left(\frac{L}{b}\right)^2 \left\{\left[12\left(\frac{m}{L}\right)^2 + 12\left(\frac{m}{L}\right) + 5\right]\left(\frac{d}{L}\right)^2 + 4\left(\frac{n}{L}\right)\left(\frac{d}{L}\right) + 4\left(\frac{n}{L}\right)^2\right\} + \left[12\left(\frac{m}{L}\right)^2 + 12\left(\frac{m}{L}\right) + 4\right]} \right\}$$
$$\tag{7.7}$$

They are formulae of nondimensional parameters, including beam's shape param-
eters $\frac{L}{b}$ and $\frac{h}{b}$, as well as beam's location parameters $\frac{d}{L}$, $\frac{m}{L}$ and $\frac{n}{L}$. Compared to C_i,
the effect of C_{bi}^* on C_{bi} is less intuitive and needs a further investigation.

7.3 Compliance Variations of Ortho-Planar Springs with Nondimensional Parameters

A symbolic formula of compliance matrix $\mathbf{C}_b$ of the ortho-planar spring is developed in Sect. 7.2 and revealed to be a 6×6 diagonal matrix. As a result, both linear and rotational compliance variations can be evaluated according to the diagonal compliance elements of $\mathbf{C}_b$. For example, the ratio of linear compliance element C_{b1} with respect to the norm of linear compliance can be written as

$$\frac{C_{b1}}{\sqrt{C_{b1}^2 + C_{b2}^2 + C_{b1}^2}} = \frac{1}{\sqrt{2(\frac{C_{b2}}{C_{b1}})^2 + 1}} \tag{7.8}$$

which is the function of linear compliance ratio $\frac{C_{b2}}{C_{b1}}$. Rotational compliance distributions can then be developed in the same way as functions of rotational compliance ratio $\frac{C_{b3}}{C_{b4}}$. Both linear and rotational compliance ratios can be written as

$$\frac{C_{b2}}{C_{b1}} = \left\{\frac{C_2}{C_3}\right\} \left\{\frac{C_{b2}^*}{C_{b1}^*}\right\}, \quad \frac{C_{b3}}{C_{b4}} = \left\{\frac{C_6}{C_5}\right\} \left\{\frac{C_{b3}^*}{C_{b4}^*}\right\} \tag{7.9}$$

It is convenient to note that the resulted compliance ratios are functions of nondimensional parameters. We are particularly interested in beam's location parameters $\frac{d}{L}$, $\frac{m}{L}$ and $\frac{n}{L}$ because they exactly reflect the influence of configuration design on the spatial compliance variations of ortho-planar springs. As a result, their effects on both linear and rotational compliance ratios are studied separately. Further, a list of compliance elements and nondimensional parameters are presented in Table 7.1. It will be helpful to set the design space for each parameter. In our study, all geometrical parameters mainly follow assumptions used in analytical modelling in Sect. 7.2, with parameter ranges referencing [1]. Specifically beam's shape parameters are fixed as $L = 18$ mm and $b = 1.6$ mm that $\frac{L}{b} \geq 10$ based on the slender beam assumption, $\frac{h}{b} \leq 1$ following design of an ortho-planar spring preferring a large out-of-plane compliance. In terms of beam's location parameters $\frac{d}{L}$, $\frac{m}{L}$ and $\frac{n}{L}$, they are determined that an ortho-planar spring can have a compact form.

Table 7.1 A list of compliance elements and nondimensional parameters used in the compliance variation analysis

Symbol	Notation
C_i	Compliance element of compliance matrix $\mathbf{C}$
C_{bi}	Compliance element of compliance matrix $\mathbf{C}_b$
C_{bi}^*	Nondimensional coefficient of compliance element C_{bi}
$\frac{L}{b}, \frac{h}{b}$	Beam's shape parameters
$\frac{d}{L}, \frac{m}{L}, \frac{n}{L}$	Beam's location parameters

7.3.1 Linear Compliance Variations with Variables $\frac{d}{L}$ and $\frac{h}{b}$

Linear compliance ratio is defined as α and can be presented by substituting Eqs. (7.1) and (7.7) in Eq. (7.9) as

$$\alpha = \frac{C_{b2}}{C_{b1}} = \left(\frac{b}{h}\right)^2 \left\{\frac{C_{b2}^*}{C_{b1}^*}\right\} \tag{7.10}$$

where C_{b1}^* and C_{b2}^* can be determined according to Eq. (7.7) as

$$C_{b1}^* = \frac{M_1 \left(\frac{d}{L}\right)^2 + \frac{4}{3}}{N_1 \left(\frac{d}{L}\right)^2 + P_1}, \quad C_{b2}^* = \left\{2\left(\frac{EI_z}{GI_x}\right)\left(\frac{d}{L}\right)^2 + \frac{2}{3}\right\} \tag{7.11}$$

where M_1, N_1 and P_1 are coefficients in C_{b1}^* and are determined when $\frac{L}{b}$ is fixed, thus C_{b1}^* is a function of $\frac{d}{L}$. C_{b2}^* is determined by coefficients $\frac{EI_z}{GI_x}$ and $\frac{d}{L}$, where $\frac{EI_z}{GI_x}$ can be calculated as

$$\frac{EI_z}{GI_x} = \frac{(1+\nu)}{2 - 1.26\left(\frac{h}{b}\right)\left(1 - \frac{h^4}{12b^4}\right)} \tag{7.12}$$

which is a function of $\frac{h}{b}$. As a result, α can be presented as a function of $\frac{h}{b}$ and $\frac{d}{L}$ as a whole. In order to evaluate the effects of $\frac{h}{b}$ and $\frac{d}{L}$, four different values of $\frac{h}{b}$ are selected from range [0.3, 1.0], and range of $\frac{d}{L}$ is set to be [0.0, 1.0] compared to the practical design space. The variations of α with respect to $\frac{h}{b}$ and $\frac{d}{L}$ are shown in Fig. 7.3, from which we can see α is extremely high when $\frac{d}{L}$ is very small and drops quickly with the increasing $\frac{d}{L}$. Then α reaches its minimum value and gradually increases again. In the preferred design space of $\frac{d}{L}$ that ranges from 0 to 0.2 [1], the minimum value of α ranges from 5 to 53, indicating ortho-planar spring significantly enlarges its out-of-plane compliance C_{b2} and minimizes its in-plane compliance C_{b1}. This comparison result agrees with previous studies [1, 2, 8] that mainly utilize the out-of-plane compliance of ortho-planar springs.

7.3.2 Rotational Compliance Variations with Variables $\frac{d}{L}$ and $\frac{h}{b}$

Following the study of linear compliance variations, rotational compliance variations are investigated in forms of $\frac{C_{b3}}{C_{b4}}$. As shown in Eq. (7.7), both C_{b3}^* and C_{b4}^* are functions of $\frac{d}{L}$, $\frac{m}{L}$ and $\frac{n}{L}$, which are more complex compared to C_{b1}^* and C_{b2}^*. As a result, it is more convenient to evaluate each nondimensional parameter's effect on

Fig. 7.3 $\alpha = \frac{C_{b2}}{C_{b1}}$ with variables $\frac{d}{L}$ and $\frac{h}{b}$, $\frac{m}{L} = -0.5$, $\frac{n}{L} = 0.5$. FEM simulation is given in Sect. 7.4

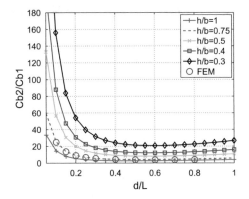

$\frac{C_{b3}}{C_{b4}}$ separately. The effect of $\frac{d}{L}$ on $\frac{C_{b3}}{C_{b4}}$ is studied first by fixing $\frac{m}{L}$ and $\frac{n}{L}$, which is symbolized as β. Substituting Eqs. (7.1) and (7.7) into Eq. (7.9), the expression of β can be written as

$$\beta = \frac{C_{b3}}{C_{b4}} = \left(\frac{b}{h}\right)^2 \left\{\frac{C_{b3}^*}{C_{b4}^*}\right\} \tag{7.13}$$

Without loss of generality, take $\frac{m}{L} = -0.5$ and $\frac{n}{L} = 0.5$, this leads to simplified expressions of C_{b3}^* and C_{b4}^* as

$$C_{b3}^* = \frac{4\left(\frac{EI_z}{GI_x}\right)\left(\frac{d}{L}\right)^2 + \frac{4}{3}}{\left[3\left(\frac{EI_z}{GI_x}\right) + 6\right]\left(\frac{d}{L}\right)^2 + 6\left(\frac{d}{L}\right) + \frac{GI_x}{EI_z} + M_2}, \quad C_{b4}^* = \frac{N_2\left(\frac{d}{L}\right)^2 + \frac{2}{3}}{P_2\left(\frac{d}{L}\right)^2 + Q_2\left(\frac{d}{L}\right) + R_2} \tag{7.14}$$

where $\frac{EI_z}{GI_x}$ has been given in Eq. (7.12) as a function of $\frac{h}{b}$. Coefficients such as M_2, N_2 and R_2 are determined by $\frac{m}{L}$ and $\frac{n}{L}$ and are constants when values $\frac{m}{L}$ and $\frac{n}{L}$ are given. Thus β can be further taken as a function of $\frac{h}{b}$ and $\frac{d}{L}$ according to Eq. (7.11). The effects of $\frac{d}{L}$ on β with different $\frac{h}{b}$ is illustrated in Fig. 7.4. Similar to α, Fig. 7.4 shows β is extremely large when $\frac{d}{L}$ is very small and drops quickly while does not increase again. When $\frac{d}{L}$ is in the range between 0 and 0.2, the minimum value of β decreases from 127 to 13 as $\frac{h}{b}$ increases from 0.3 to 1. These comparison results suggest ortho-planar spring has a larger bending compliance C_{b3} compared to torsional compliance C_{b4}. When each limb becomes thicker, even when $\frac{h}{b} = 1$, the minimum value of β is still as big as 13.

Fig. 7.4 $\beta = \frac{C_{b3}}{C_{b4}}$ with variables $\frac{d}{L}$ and $\frac{h}{b}$, $\frac{m}{L} = -0.5$, $\frac{n}{L} = 0.5$. FEM simulation is given in Sect. 7.4

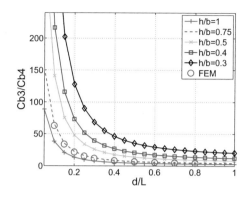

7.3.3 Rotational Compliance Variations with Variables $\frac{m}{L}$ and $\frac{h}{b}$

The effect of $\frac{m}{L}$ on $\frac{C_{b3}}{C_{b4}}$ is studied following $\frac{d}{L}$ and is symbolized as γ. By fixing $\frac{d}{L} = 0.2$ and $\frac{n}{L} = 0.5$ and submitting them in Eq. (7.7), γ can be simplified as a function of $\frac{h}{b}$ and $\frac{m}{L}$ as

$$C_{b3}^* = \frac{0.16\left(\frac{EI_z}{GI_x}\right) + 1.33}{12\left(\frac{m}{L}\right)^2 + 12\left(\frac{m}{L}\right) + \left[0.12\left(\frac{EI_z}{GI_x}\right) + \frac{GI_x}{EI_z} + M_3\right]},$$

$$C_{b4}^* = \frac{N_3}{P_3\left(\frac{m}{L}\right)^2 + Q_3\left(\frac{m}{L}\right) + R_3} \qquad (7.15)$$

where coefficients M_3, N_3 and R_3 are determined when values $\frac{L}{b}$, $\frac{d}{L}$ and $\frac{n}{L}$ are given. The variance of γ with two variables is illustrated in Fig. 7.5, where $\frac{h}{b}$ ranges from 0.3 to 1, and $\frac{m}{L}$ ranges from -1 to 1. From Fig. 7.5 we can see γ increases from the beginning and reaches the maximum value when $\frac{m}{L} = -0.5$. This inspiring result can also be calculated analytically according to Eq. (7.7). Similar to β, γ decreases with increase of $\frac{h}{b}$. When $\frac{h}{b}$ increases from 0.3 to 1, the maximum value of γ decreases from 127 to 13. This also agrees with the variation of β provided in Fig. 7.4.

7.3.4 Rotational Compliance Variations with Variables $\frac{n}{L}$ and $\frac{h}{b}$

Finally the effect of $\frac{n}{L}$ on the ratio $\frac{C_{b3}}{C_{b4}}$ symbolized as δ is investigated. Following the same simplification procedure above, we fix $\frac{d}{L} = 0.2$ and $\frac{m}{L} = 0.4$ and other design

Fig. 7.5 $\gamma = \frac{C_{b3}}{C_{b4}}$ with variable $\frac{m}{L}$ and $\frac{h}{b}$, $\frac{d}{L} = 0.2$, $\frac{n}{L} = 0.5$. FEM simulation is given in Sect. 7.4

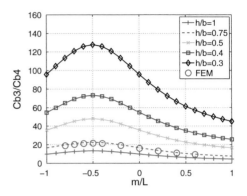

parameters and substitute them in Eq. (7.7), which results in simplified expressions of C_{b3}^* and C_{b4}^* as

$$C_{b3}^* = \frac{0.16\left(\frac{EI_z}{GI_x}\right) + 1.33}{12\left(\frac{n}{L}\right)^2 + 2.4\left(\frac{n}{L}\right) + \left[0.12\left(\frac{EI_z}{GI_x}\right) + \frac{GI_x}{EI_z} + M_4\right]},$$

$$C_{b4}^* = \frac{N_4}{P_4\left(\frac{n}{L}\right)^2 + Q_4\left(\frac{n}{L}\right) + R_4} \tag{7.16}$$

Similarly, values of coefficients M_4, N_4 and R_4 are determined by $\frac{L}{b}$, $\frac{d}{L}$ and $\frac{m}{L}$. As a result, δ is a function of variables $\frac{h}{b}$ and $\frac{n}{L}$. The diagram of δ with change of $\frac{h}{b}$ and $\frac{n}{L}$ is illustrated in Fig. 7.6, where $\frac{h}{b}$ ranges from 0.3 to 1, and $\frac{n}{L}$ ranges from -1 to 1. Figure 7.6 demonstrates an opposite tendency of $\frac{C_{b3}}{C_{b4}}$ compared to that revealed in Fig. 7.5, as δ decreases to its minimum value with increase of $\frac{n}{L}$ and then increases again regarding each $\frac{h}{b}$. $\frac{n}{L} = -0.1$ corresponds to the minimum value of δ in this comparison. Generally it equals to $-\frac{d}{2L}$ according to Eq. (7.7) and is not a constant value. In addition, the minimum value of δ decreases from 22 to 2 as $\frac{h}{b}$ increases from 0.3 to 1. This result suggests the torsional compliance C_{b4} is closer to the bending compliance C_{b3} compared to the results obtained in Fig. 7.5.

In summary, both linear and rotational compliance variations of ortho-planar springs are investigated using nondimensional parameters. It is revealed for the first time that ortho-planar spring not only has a large out-of-plane linear compliance C_{b2}, but also has a large bending compliance C_{b3}. This can be identified from Figs. 7.3 and 7.4. Then the effect of limb's location parameters on planar spring's rotational compliances is evaluated and compared in Figs. 7.5 and 7.6, leading to new findings of $\frac{m}{L} = -0.5$ and $\frac{n}{L} = -\frac{d}{2L}$ resulting in the maximum and minimum values of $\frac{C_{b3}}{C_{b4}}$, as can be seen in Fig. 7.7. They correspond to two typical types of ortho-planar springs [2], including the side-type and radial-type. Side-type planar spring is obtained when $\frac{m}{L} = -0.5$, and radial-type corresponds to $\frac{n}{L} = -\frac{d}{2L}$.

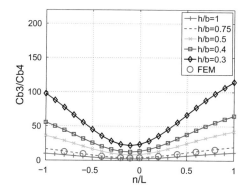

Fig. 7.6 $\delta = \frac{C_{b3}}{C_{b4}}$ with variable $\frac{n}{L}$ and $\frac{h}{b}$, $\frac{d}{L} = 0.2$, $\frac{m}{L} = 0.4$. FEM simulation is given in Sect. 7.4

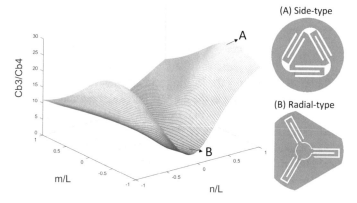

Fig. 7.7   Variations of $\frac{C_{b3}}{C_{b4}}$ with variables $\frac{m}{L}$ and $\frac{n}{L}$

7.4   Compliance Validation of Ortho-Planar Springs Using a Multi-body Based FEA Simultion Approach

In this section, ortho-planar springs with various configurations are designed and their spatial compliance performances are evaluated using FEM simulation. It is noticed that some assumptions are made in analytical modeling in Sect. 7.2, such as the base and functional platform are assumed to be rigid. In accordance with this, a multi-body based simulation approach is adopted, where slender beams are modelled as flexible parts and other segments as rigid parts. As a result, only flexible beams are allowed to deform during a loading process. Figure 7.8 shows the multi-body based simulation results of a side-type ortho-planar spring under two different types of load. FEM simulation is conducted in the ANSYS static structural analysis environment. Material properties follow the specifications of VisiJet EX200 Plastic used in 3D printing in order to be consistent with the following experiment tests and

Fig. 7.8 Finite element simulations using a multi-body based modeling approach

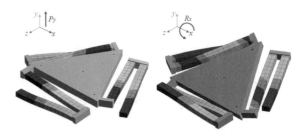

Table 7.2 Simulated planar-spring configurations with variable geometrical parameters

Group	Fixed parameter (mm)	Variable parameter	Value range
A	$m = -9, n = 9$	$\frac{d}{L}$	[0.1, 0.8]
B	$m = -9, n = 9$	$\frac{d}{L}$	[0.1, 0.8]
C	$d = 3.6, n = 9$	$\frac{m}{L}$	[−0.8, 0.6]
D	$d = 3.6, m = 7.2$	$\frac{n}{L}$	[−0.8, 0.8]

are listed in Table 7.3. Regarding the loading and boundary conditions, the end of the outside beam in each limb is fixed to the ground, and loads are added at the functional platform with their maximum values determined that the resulted maximum stresses are not exceeding the tensile strength of the material.

Subsequently a total number of 30 types of ortho-planar springs are selected and simulated using this multi-body based modeling approach. Their configurations are determined following the analytical compliance studies in Sect. 7.3. In particular, four groups of planar-spring models are designed. Models in group A and B are designed according to Sects. 7.3.1 and 7.3.2 that evaluate the effects of $\frac{d}{L}$ on $\frac{C_{b2}}{C_{b1}}$ and $\frac{C_{b3}}{C_{b4}}$. Groups C and D contain models according to Sects. 7.3.3 and 7.3.4 to address the effects of $\frac{m}{L}$ and $\frac{n}{L}$ on $\frac{C_{b3}}{C_{b4}}$. They have the same beam-shape parameters as $L = 18$ mm, $b = 1.6$ mm and $h = 1.2$ mm. The beam's location parameters are listed in Table 7.2. In addition, a complete list of their corresponding geometrical configurations are presented in Fig. C.1 in Appendix C.

In order to estimate compliance elements $C_{bi} (i = 1, \ldots, 4)$, their corresponding load types are applied, including lateral force, linear out-of-plane force, bending moment and torsional moment. The resulted deformations are recorded and used to obtain compliance ratios $\frac{C_{b2}}{C_{b1}}$ and $\frac{C_{b3}}{C_{b4}}$. Then they are compared with analytical models in Figs. 7.3, 7.4, 7.5 and 7.6. The comparison results indicate FEM simulations are close to related analytical values with all selected planar-spring models. Further their discrepancies are calculated and presented in Fig. 7.9. Red line highlighted by small circles represents the discrepancies of ratio $\frac{C_{b2}}{C_{b1}}$ with variable $\frac{d}{L}$, which corresponds to planar-spring configurations in Group A. Green-cross line represents

Fig. 7.9 Discrepancies between FEA simulations and analytical models in accordance with their comparison results shown Fig. 7.3, 7.4, 7.5 and 7.6. Test points correspond to planar-spring configurations listed in Fig. C.1

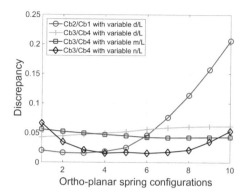

the discrepancies of ratio $\frac{C_{b3}}{C_{b4}}$ with variable $\frac{d}{L}$, which corresponds to planar-spring configurations in Group B. The rest are the same.

From Fig. 7.9 we can see the discrepancy values of most configurations remain as low as 5%. It is noticed that the discrepancies of last four test configurations in Group A increase significantly, which correspond to planar springs with $\frac{d}{L}$ that ranges from 0.5 to 0.8. However, in the practical designs where $\frac{d}{L} \leq 0.2$ [1], even with a larger range in our case, their discrepancies still remain lower than 5%. In addition, all curves remain at a low-discrepancy level in terms of the rotational compliance ratio comparisons, which suggests the developed analytical model is able to predict rotational compliances of planar springs well with a large range of configuration space. In summary, the configurations listed in Fig. C.1 can help us visualize the geometrical shapes of different planar spring designs, and the corresponding analytical models enable us to predict their spatial compliance performances with a good accuracy. For instance, we can easily identify some configurations listed in Fig. C.1 are not common from a practical-design point of view [1], such as configurations with variable $\frac{m}{L} = 0.6$ and $\frac{n}{L} = 0.8$, etc. They may require space but their spatial compliance performances are not optimal; on the other hand, the typical side-type and radial-type planar springs are good design options not only because of their relatively compact form, but also their better spatial compliance performances.

7.5   Experiment Evaluation of Linear Compliance C_{b2} of Ortho-Planar Springs

In this section, the widely used linear compliance C_{b2} of ortho-planar springs are evaluated through experiments. An experiment platform was built that can apply displacement load at planar-spring samples and obtain the reaction force. Both displacement load and reaction force are recorded to develop displacement-force characteristics of tested samples.

Experiment samples are presented in Fig. 7.10, with both side-type and radial-type planar springs. They are manufactured using 3D rapid-prototyping machine ProJet HD 3000, the material of which is VisiJet EX200 Plastic. Material specifications and geometrical parameters of tested samples are listed in Table 7.3. Apart from key design parameters, the shape of the base and functional platform, as well as the size of connecting parts are carefully designed to strengthen the structure and eliminate unnecessary deformation, such as lateral deformation, bending of the base and functional platform, etc. Screwing holes are made to connect either bases or functional platforms to the testing platform. Red markers in Fig. 7.10 are the contact points. The red solid-diamond markers in the center are testing points for linear compliance C_{b2}, while red solid-circle markers around the center points are contact points for testing coupling compliance of C_{b2} and C_{b3}.

The experiment setup for testing linear compliance C_{b2} is presented in Fig. 7.11, where the base of a sample was fixed to a support pole that was assembled on a linear guide. A Nano 17 Six-DOF force/torque sensor assembled on a moving block was used to record the reaction force data. The moving block can move along the linear guide, it had two degrees of freedom in the plane that is perpendicular to the motion direction of linear guide. As a result, the sensor tip can be adjusted to attach to the center of planar spring at the initial position. The linear guide was driven by a stepper motor and it was connected to a positioning controller. A LABVIEW

Fig. 7.10 Experiment samples of both side-type and radial-type planar springs

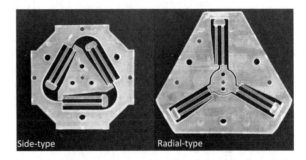

Table 7.3 Material properties and geometrical dimensions of ortho-planar spring samples

Material property	Density (g cm^{-3})	Young's modulus (MPa)	Possion's ratio	Tensile strength (MPa)
	1.02	1159	0.42	42.4
Beam's shape parameters	L (mm)	b (mm)	h (mm)	d (mm)
	18	1.2	1.2	3
Beam's location parameters	m (mm)	n (mm)		
Side-type	−9	9		
Radial-type	7.2	0		

Fig. 7.11 Experiment setup
to evaluate the linear
compliance C_{b2}

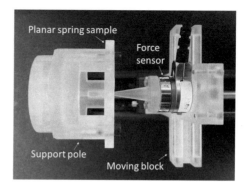

program was developed to control the linear guide with the desired speed and record
the reaction force data from Nano 17 force/torque sensor with the same sample rate.
The experiment reported here was performed at 25 °C in temperature and 50% in
relative humidity. Then experiments were conducted by adding displacement load
at the center of planar-spring sample and recording the reaction force. The forward
and backward speed of the moving block was set to be $v = 5\,\text{mm/min}$, which was
slow enough to decrease the elastic hysteresis, and eight loading displacements were
selected every 0.4 mm from 0.4 to 3.2 mm so the resulted maximum stress is well
below the tensile strength of the material. Each specimen was tested with ten experiment runs, the last five of them were selected to obtain the average value of reaction
force and standard deviation at each tested point.

Then experiments were conducted by adding displacement load at the center of
planar-spring sample and recording the reaction force. The forward and backward
speed of the moving block was set to be $v = 5\,\text{mm/min}$, which was slow enough to
decrease the elastic hysteresis, and eight loading displacements were selected every
0.4 mm from 0.4 to 3.2 mm so the resulted maximum stress is well below the tensile
strength of the material. Each specimen was tested with ten experiment runs, the last
five of them were selected to obtain the average value of reaction force and standard
deviation at each tested point.

The experiment results are then compared with analytical modelings. Analytical
values of C_{b2} are calculated by substituting design parameters listed in Table 7.3
into Eqs. (7.6) and (7.7). The comparison results are shown in Fig. 7.12, where the
black solid-circle line represents analytical models (Side-type and Radial-type have
the same C_{b2}), the red dash-square line is the experiment result of side-type spring
and red dash-circle one is that of radial-type spring. Error bars represent standard
deviations of every tested point.

Comparison results suggest experiment results match the analytical models well
for both side-type and radial-type planar springs. Reaction forces of experimental
tests are larger than analytical values at the beginning and converging towards them

Fig. 7.12 Displacement-
force comparison results of
compliance $(C_b)_2$, black
solid-circle line represents
analytical model, red solid
line represents FEA
simulation results, and blue
dash lines represent
experiment results

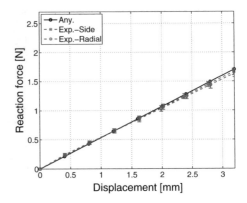

with the increase of displacement load. In terms of side-type planar spring, the average
discrepancy between experimental reaction forces and analytical results is about
2.7%, which is 4% in the radial-type planar springs. These remain at a satisfactory
level considering the fragile properties of 3D printing material. Further, C_{b2} of both
side-type and radial-type planar springs are compared. It can be seen from Fig. 7.12
that their experiment curves are most overlapping with each other. It is noticed that
some discrepancies occur between their experiment curves, which are due to the fact
the connecting parts in side-type spring samples are stronger than those in radial-
type spring samples. This can also be observed from the sample diagrams shown in
Fig. 7.10. Despite that, the maximum discrepancy between them is as small as 1.2%,
which indicates these two types of planar springs have the same linear compliance
C_{b2}.

7.6   Experiment Evaluation of Coupling Compliance C_{b2} and C_{b3} of Ortho-Planar Springs

Following evaluation of linear compliance C_{b2}, the coupling compliance performance
between C_{b2} and C_{b3} are examined in this section. The experiment setup is the same
as that used in Sect. 7.5 but with a different testing principle. In this experiment,
planar spring samples are pushed at contact points on the out-layer other than the
center functional platform. They are labelled with red solid-circle markers shown in
Fig. 7.10. Accordingly mathematical modeling of this experiment is firstly introduced
in Sect. 7.6.1, followed by experiment description and evaluation in Sect. 7.6.2.

7.6.1 Mathematical Modeling of Coupling Compliance C_{b2} and C_{b3}

The experiment setup was designed to address both C_{b2} and C_{b3} using displacement load and reaction force. The related mathematical diagram is shown in Fig. 7.13, where the functional platform of a planar spring is fixed to a support pole (which is hidden) and the outer layer is pushed with a displacement load L at the touching point A. The force arm is d_0 and remains constant during the whole manipulation process. F is the reaction force in the y-axis direction, which results in a force in the same direction and a torque about the z-axis regarding the origin of coordinate frame $\{O, x, y, z\}$. The equivalent force generates a shift distance L_1 and the equivalent torque generates a rotation angle θ that corresponds to a distance L_2. The sum of L_1 and L_2 equals to the overall stroke L. The relationship between L and F can be formulated as

$$L = L_1 + d_0 \tan \theta$$
$$= C_{b2} F + d_0 \tan (C_{b3} F d_0) \tag{7.17}$$

Thus the combined compliance performance of C_{b2} and C_{b3} can be determined according to Eq. (7.7) by knowing L and F. Further FEA simulation is utilized to obtain the center shift distance L_1 and rotation angle θ. The boundary and load conditions applied at the model are the same as those used in the mathematical model presented in Fig. 7.13, where the functional platform was fixed and a remote displacement L along the y axis was applied at the point A with a distance of $d_0 = 21$ mm. Then reaction forces corresponding to different stroke L were recorded. In addition, the rotation angle θ and center-shift-distance L_1 were extracted from the simulation results. According to Eq. (7.17), reaction force F_l resulted from L_1 was calculated using $\frac{L_1}{C_{b2}}$, and force F_θ resulted from θ was calculated using $\frac{\theta}{C_{b3}d_0}$. Then their average values were calculated to develop the analytical displacement-force curves.

Fig. 7.13 Mathematical model used to evaluate the coupling compliance performance of C_{b2} and C_{b3}

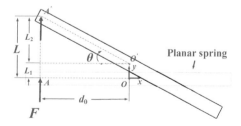

7.6.2 Experiment Validation of Coupling Compliance C_{b2} and C_{b3}

Subsequently, experiments were conducted to validate the coupling performances of C_{b2} and C_{b3}. The tested samples are the same as those used in the first experiment, but the center of spring sample is fixed and the out layer is pushed instead in this test. The corresponding experiment setup and the deformation of planar-spring samples are illustrated in Fig. 7.14. The moving block changed its position so that the force-sensor tip could reach the designed contact points. The forward and backward speed of the moving block was set to be $v = 5$ mm/min, and eight loading displacements were selected every 0.5 mm from 0.5 to 4 mm. Further, for both types of ortho-planar springs, each sample was pushed at three different contact points shown in Fig. 7.10, which are located at the same circle with a radius of $d_0 = 21$ mm. By comparing reaction forces at different contact points, we are able to examine if ortho-planar spring is symmetric and has the same bending compliance about different axes.

The displacement-reaction force curves were developed and compared with analytical values. The comparison results are shown in Fig. 7.15. The black solid lines represent analytical results and red dash lines are experiment results. Error bars represent standard deviations of tested points. Overall experiment results match the analytical models with good accuracy. In detail, the tested reaction forces increase with a slightly quicker rate and then become closer to the analytical values, which is similar to the observation obtained in Fig. 7.12. In terms of the side-type planar spring, the average discrepancy between the experimental value and the analytical model is 6%, which is 3.8% in the radial-type planar spring. The agreements between analytical models and experimental results validate the mathematical model developed in Sect. 7.6.1, which suggests the deformation of the planar spring structure can be decoupled into a linear displacement and a rotation about the center of the functional platform. This can also be identified from the experimental observation shown in Fig. 7.14.

Further, the symmetrical property of planar spring samples can be evaluated by comparing the reaction forces at different contact points. Here the coefficient of variance (COV) is used. In terms of side-type spring samples, the average COV of different testing points is as small as 3.7%, which is 3.6% in the radial-type

Fig. 7.14 Deformation of a planar-spring sample during the displacement-force experiment test

Fig. 7.15 Displacement-
force Comparison results of
coupling compliance $(C_b)_2$
and $(C_b)_4$, black solid lines
represent analytical models,
red solid lines represents
FEA simulation curves, and
blue dash lines represent
experiment results. Error
bars represent standard
deviations at corresponding
tested points

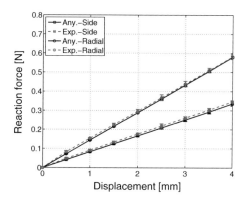

spring samples. Also two-sample t tests are applied to examine their correlations.
The hypothetical test result $h = 0$ indicates a failure to reject null hypothesis at 5%
significance level, which suggests each planar spring sample has the same coupling
compliance performance at different contact points.

7.7   A Continuum Manipulator Based on the Serial Integration of Ortho-Planar-Spring Modules

Based on the FEM simulations and experiment validations, the analytical approach
is revealed to be accurate in modelling ortho-planar spring's compliance behaviours.
Also, comparison results suggest the ortho-planar spring demonstrates a large bending compliance in comparison to torsional compliance, indicating its potential value
in developing continuum manipulators that utilize the bending compliance of individual segments [25–27]. To explore this potential application, one physical prototype
of a continuum structure with serial chains of ortho-planar springs is developed. Particularly the side-limb type is selected due to its compact form as well as its optimal
spatial compliance performance which provides the maximum ratio of bending compliance to torsional compliance. One segment is shown in the left part of Fig. 7.16,
which is made of acrylonitrile butadiene styrene (ABS) plastic using a rapid prototyping 3-D printer. This segment presents a very compact form as it has a diameter of
18 mm. The segment contains two layers, the upper layer is the ortho-planar spring
that provides the desired deformation, and the lower layer is a connector that is used
to assemble another segment. The distance between two layers is designed properly
so that the ortho-planar spring can rotate freely within the gap between two layers
without static failure. The distance value can be determined based on the maximum
workspace of the ortho-planar spring using the analytical model developed, in this
prototype, the maximum bending angle is set as $13°$ only for a purpose of demonstration.

Fig. 7.16 Prototype of a
continuum manipulator
based on the serial
integration of ortho-planar
springs

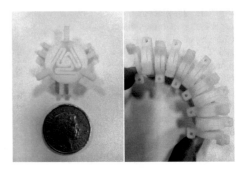

Subsequently, the continuum structure was developed by assembling ortho-planar spring segments in a serial chain. The prototype is illustrated in the right part of Fig. 7.16, which consists of ten segments. It is bent at two ends, the maximum bending angle reaches almost 130° with a longitude length of 70 mm, which demonstrates a good bending capability of this continuum structure. In [28, 29] the continuum robots utilized elastic backbone to generate bending deformation, which requires the backbone to be sufficiently long and thin. In comparison with the thin backbone design, planar-spring based segment demonstrates concentrated compliance and larger flexibility compared to its length. Also, the compliance matrix of ortho-planar spring can be designed and calculated accurately within a large deformation range based on the analytical model provided in this paper. Another benefit is that the compliance matrix is diagonal, this avoids center shift problem [30, 31] and guarantees the whole continuum manipulator to deform along its central axis exactly during the manipulation process.

7.8   Conclusions

This chapter provided a comprehensive study of ortho-planar springs from analytical modelling to finite-element-simulation and experiment validation, leading to novel findings and applications of the spatial compliance characteristics of ortho-planar springs. In the analytical modelling, the symbolic formula of an ortho-planar spring's compliance matrix was developed in its diagonal form. Based on the compliance matrix, compliance variations of ortho-planar springs with nondimensional parameters were examined, revealing an ortho-planar spring not only demonstrating a large out-of-plane linear compliance, but also having a large bending compliance. In particular, the rotational compliance studies led to synthesis of typical side-type and radial-type ortho-planar springs, which have the same linear compliances but significantly different rotational compliances.

Finite element (FEM) simulations of a total number of 30 types of planar-spring models were conducted and compared with their analytical results. A multi-body based simulation approach was adopted to conduct the FEM simulation. The comparison results, as well as the subsequent discrepancy analysis, indicate the developed analytical model can predict the spatial compliance performance of an ortho-planar spring accurately within the practical design space of ortho-planar springs.

Experiments of the typical side-limb and radial-limb planar springs were then carried out with respect to the linear out-of-plane compliance and the discovered rotational bending compliance. As a result, experiment results of both compliances matched the analytical models well. Further, the large bending characteristics of the ortho-planar spring were utilized to build a module-based continuum manipulator. Compared with other designs, this manipulator not only demonstrates a large bending compliance but also has a center-of-compliance in each segment which avoids center-shift in manipulation. These advantages indicate the potential value of the ortho-planar spring in the design of continuum manipulators in bio-robotics and medical-robotics research fields.

References

1. Howell, L.L.: Compliant Mechanisms. Wiley-Interscience (2001)
2. Parise, J.J., Howell, L.L., Magleby, S.P.: Ortho-planar linear-motion springs. Mech. Mach. Theory **36**(11), 1281–1299 (2001)
3. Howell, L.L., Thomson, S., Briscoe, J.A., Parise, J.J., Lorenc, S., Larsen, J.B., Huffmire, C.R., Burnside, N., Gomm, T.A.: Compliant, ortho-planar, linear motion spring, Jan. 10, 2006. US Patent 6,983,924
4. Whiting, M.J., Howell, L.L., Todd, R.H., Magleby, S.P., Anderson, M.C., Rasmussen, N.O.: Continuously variable transmission or clutch with ortho-planar compliant mechanism (2008). US Patent 7,338,398
5. Jacobsen, J.O., Winder, B.G., Howell, L.L., Magleby, S.P.: Lamina emergent mechanisms and their basic elements. J. Mech. Robot. **2**(1), 011003 (2010)
6. Nguyen, N.-T., Truong, T.-Q., Wong, K.-K., Ho, S.-S., Low, C.L.-N.: Micro check valves for integration into polymeric microfluidic devices. J. Micromech. Microeng. **14**(1), 69 (2004)
7. Smal, O., Dehez, B., Raucent, B., De Volder, M., Peirs, J., Reynaerts, D., Ceyssens, F., Coosemans, J., Puers, R.: Modelling, characterization and testing of an ortho-planar micro-valve. J. Micro-Nano Mechatron. **4**(3), 131–143 (2008)
8. Ataollahi, A., Fallah, A.S., Seneviratne, L.D., Dasgupta, P., Althoefer, K.: Novel force sensing approach employing prismatic-tip optical fiber inside an orthoplanar spring structure.
9. Dimentberg, F.M.: The screw calculus and its applications in mechanics. Technical report, DTIC Document (1968)
10. Loncaric, J.: Geometrical analysis of compliant mechanisms in robotics (euclidean group, elastic systems, generalized springs (1985)
11. Patterson, T., Lipkin, H.: Structure of robot compliance. Trans. Am. Soc. Mech. Eng. J. Mech. Des. **115**, 576–576 (1993)
12. Dai, J.S., Kerr, D.: A six-component contact force measurement device based on the Stewart platform. Proc. Inst. Mech. Eng. Part C: J. Mech. Eng. Sci. **214**(5), 687–697 (2000)
13. Ciblak, N., Lipkin, H.: Design and analysis of remote center of compliance structures. J. Robot. Syst. **20**(8), 415–427 (2003)
14. Su, H.-J., Shi, H., Yu, J.: A symbolic formulation for analytical compliance analysis and synthesis of flexure mechanisms. J. Mech. Des. **134**, 051009 (2012)
15. Gosselin, C.: Stiffness mapping for parallel manipulators. IEEE Trans. Robot. Autom. **6**(3), 377–382 (1990)

16. Schotborgh, W.O., Kokkeler, F.G., Tragter, H., van Houten, F.J.: Dimensionless design graphs for flexure elements and a comparison between three flexure elements. Precis. Eng. **29**(1), 41–47 (2005)
17. Dai, J.S., Xilun, D.: Compliance analysis of a three-legged rigidly-connected platform device. J. Mech. Des. **128**(4), 755–764 (2006)
18. Ding, X., Dai, J.S.: Characteristic equation-based dynamics analysis of vibratory bowl feeders with three spatial compliant legs. IEEE Trans. Autom. Sci. Eng. **5**(1), 164–175 (2008)
19. Pashkevich, A., Chablat, D., Wenger, P.: Stiffness analysis of overconstrained parallel manipulators. Mech. Mach. Theory **44**(5), 966–982 (2009)
20. Klimchik, A., Pashkevich, A., Chablat, D.: Cad-based approach for identification of elastostatic parameters of robotic manipulators. Finite Elem. Anal. Des. **75**, 19–30 (2013)
21. Young, W.C., Budynas, R.G.: Roark's Formulas for Stress and Strain, vol. 6. McGraw-Hill, New York (2002)
22. Dai, J.S.: Geometrical Foundations and Screw Algebra for Mechanisms and Robotics. Higher Education Press, Beijing (2014). ISBN 9787040334838. (translated from Dai, J.S.: Screw Algebra and Kinematic Approaches for Mechanisms and Robotics. Springer, London (2016))
23. Dai, J.S.: Screw Algebra and Kinematic Approaches for Mechanisms and Robotics. Springer, London (2016)
24. Murray, R.M., Li, Z., Sastry, S.S., Sastry, S.S.: A Mathematical Introduction to Robotic Manipulation. CRC Press (1994)
25. Webster, R.J., Jones, B.A.: Design and kinematic modeling of constant curvature continuum robots: a review. Int. J. Robot. Res. **29**(13), 1661–1683 (2010)
26. Qi, P., Qiu, C., Liu, H.B., Dai, J.S., Seneviratne, L., Althoefer, K.: A novel continuum-style robot with multilayer compliant modules. In: Proceedings of IEEE/RSJ International Conference on Intelligent Robots and Systems, Chicago, Illinois, USA, September 14–18, 2014. IEEE, Chicago (2014)
27. Qi, P., Qiu, C., Liu, H., Dai, J., Seneviratne, L., Althoefer, K.: A novel continuum manipulator design using serially connected double-layer planar springs. IEEE/ASME Trans. Mechatron. **99**, 1–1 (2015)
28. Gravagne, I.A., Walker, I.D.: On the kinematics of remotely-actuated continuum robots. In: 2000 Proceedings of the IEEE International Conference on Robotics and Automation, ICRA'00, vol. 3, pp. 2544–2550. IEEE (2000)
29. Gravagne, I.A., Rahn, C.D., Walker, I.D.: Large deflection dynamics and control for planar continuum robots. IEEE/ASME Trans. Mechatron. **8**(2), 299–307 (2003)
30. Xu, P., Jingjun, Y., Guanghua, Z., Shusheng, B.: The stiffness model of leaf-type isosceles-trapezoidal flexural pivots. J. Mech. Des. **130**(8), 082303 (2008)
31. Martin, J., Robert, M.: Novel flexible pivot with large angular range and small center shift to be integrated into a bio-inspired robotic hand. J. Intell. Mater. Syst. Struct. **22**(13), 1431–1437 (2011)

Chapter 8
Large Deformation Analysis of Compliant Parallel Mechanisms

8.1 Introduction

Origami is an art from China and Japan which folds a flat sheet of paper into a 3D object with various shapes [1, 2]. With the growing interest from both academy and industry, the knowledge of origami in artistic discipline is being integrated into a wide range of research and engineering applications. Dai and Jones [3] and Dai and Caldwell [4] used origami to investigate complex decorative cartons with their versatility in a design in the packaging industry. Howell [5] discovered the link between origami and compliant mechanisms and used it to generate a new concept for nano-manufacture [6] and to develop novel compliant mechanical systems [7]. Song et al. [8] and Ma and You [9] utilized origami patterns to design thin-walled structures for energy absorption. For the thick-wall problem, Zirbel et al. [10] and Chen et al. [11] addressed it in the origami folding process. Under axial loading, these structures are able to deform following origami patterns and smooth crushing process. Further, with the rapid development of 2D fabrication and micro-actuation technologies [12, 13], passive origami structures have now been turned into active robots, such as a deformable wheel robot by Lee et al. [14], continuum manipulators by Onal et al. [15], Vander Hoff et al. [16] and a robot end-effector by Zhang et al. [17].

Among these applications, foldability is the primary concern of most developed origami structures. Foldability of an origami structure represents its ability to deform after it is folded, which is mainly determined by the motion of its crease pattern [18, 19]. The design of crease-pattern draws a substantial amount of attention in the study of origami folding geometry [20], its folding trajectory [21] and computational efficiency [2, 22]. These developments focus on finding mathematical rules of folding 3D origami structures from a single piece of paper. In terms of motion study of origami structures, a widely used method is the mechanism-equivalent approach [18], which is able to model origami structures and extend them to create mechanism equivalents such as the palm-foldable hand [23] and innovative mechanisms [24–28]. In this

C. Qiu and J. S. Dai, *Analysis and Synthesis of Compliant Parallel Mechanisms—Screw Theory Approach*, Springer Tracts in Advanced Robotics 139, https://doi.org/10.1007/978-3-030-48313-5_8

approach, an origami structure can be mapped to its mechanism equivalent by treating creases as revolute joints and panels as links, and its motion analysis can be achieved using the well-developed methodology in kinematics.

In using this mechanism-equivalent approach, elasticity that exists in creases introduces the coherent stiffness of developed 3D origami structures and coexists with their foldabilities. The stiffness characteristics of origami structures have been proved to be essential in many practical applications. For instance, stiffness of crease patterns determines the energy that can be absorbed in the thin-walled energy absorbers [8, 9] and that affects the actuation force as well as the payload for a given structural configuration in origami continuum manipulators [15–17].

Stiffness of an origami structure is naturally related to its force behaviour. Related research include folding-force evaluation of single origami creases [29–32] as well as the whole origami structures [33–38]. As such, a systematic approach is needed to be implemented. This requires understanding the intrinsic force-transmission principle and then integrates the elastic characteristics of single creases into the overall force performance of origami structures. A thorough observation reveals when an origami structure is deformed, resistive torques from folded creases result in an integrated reaction force to counter the external load applied to an origami structure. This inspires us to calculate the reaction force of origami structures using the mechanism-equivalent approach, and take resistive torques into account using the force analysis of the equivalent mechanisms.

As a result, this chapter extends the mechanism-equivalent approach by treating origami structures as redundantly actuated mechanisms and develops a general approach in modelling the reaction force of an origami structure according to the forward-force analysis of its equivalent mechanism. In particular, the concept of repelling screws is used for the first time in this modelling approach. Section 8.2 presents a complete experimental test of single creases of origami folds, which reveals an origami crease can be treated as a one-DOF flexible element with embedded torsional stiffness. Section 8.3 introduces the theoretical background of force analysis of origami structures in the framework of screw theory, with the repelling screw and the resistive torques of creases. The analytical modeling is followed by an algebraic calculation process in Sect. 8.4, which calculates the reaction force of the widely used origami waterbomb base. This leads to a force modeling of a waterbomb-base integrated parallel structure in Sect. 8.5 and its finite-element-simulation validation in Sect. 8.6 respectively.

8.2   Stiffness Characteristics of Single Creases

In this section, we explore the individual crease's stiffness characteristics with folding experiments. As has been discussed in Sect. 1.1, a single crease in an origami fold is a multi-layer type flexible element that has one rotational degree of freedom. A typical crease-type flexure is shown in Fig. 8.1. It consists of multiple layers, when

Fig. 8.1 A multi-layer
crease-type flexible element

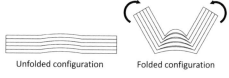

Unfolded configuration   Folded configuration

Fig. 8.2 Experiment setup

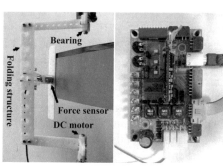

it is folded, the outside layers are pulled and inside layers are pushed. Together they
provide the inherent torsional stiffness to resist the folding motion of the crease.

Then a series of experiments were conducted to evaluate the effects of different
parameters on the stiffness properties of a single crease [39], including the folding
cycle, crease length as well as folding velocity. The crease samples used in the
experiments are made from an open flap of FLUORESCENT CARDS ORANGE.

The experiment is set up to test the motion-moment characteristics of single
creases. As shown in Fig. 8.2, the main body of an origami carton was fixed on a base
and a folding structure was designed to fold the carton panel. The folding structure
had two supporting beams, the upper one was connected to a rotation bearing, and the
lower one was connected to a DC motor. A 6-axis force/torque sensor (ATI nano17,
sample rate = 1000 Hz) fixed on the folding structure was used to record folding
force generated.

During the folding process, both upper bearing axis and lower motor axis were
collinear to the rotation axis of the crease. This ensures the z-axis of the sensor is
perpendicular to the manipulated panel, such that z-axis force can be extracted as
normal folding force. Accordingly, the folding moment was calculated by multiplying
the sensor's z-axis force and vertical distance between rotation axis and sensor's
location. The DC motor was connected to a positioning controller embedded in
a LABVIEW program and was driven with required rotation angles and rotation
velocities. The experiment reported here was performed at 25 °C in temperature and
50% in relative humidity.

8.2.1 Virgin and Repeated Folding-Stiffness Characteristics

The virgin and repeated folding moment characteristics of a single crease are studied in this section. The length of tested creases was $l = 0.05$ m and folding angular velocity was chosen as $\omega = 1$ rpm with folding angle ranging from $0°$ to $90°$. Ten crease samples were made from an open blank of FLUORESCENT CARDS that has thickness $h = 0.35 \times 10^{-3}$ m, they were pre-creased with creasing mould so that panels can be folded precisely along predetermined lines. Each specimen was tested with ten experiment runs, where the first experiment run was chosen as the virgin folding manipulation, and experiment runs from three to ten were defined as repeated folding manipulations.

Figure 8.3 shows the folding moment comparisons between virgin and repeated folding manipulations, where the dashed lines represent mean values of virgin folding moment curves, solid lines represent mean values of repeated folding moment curves, and the dash-dot lines represent $\pm$ one standard deviation. From Fig. 8.3 we can see the virgin folding curve follows a significantly different trend than the repeated folding curve. The slope of virgin folding curve increases at a higher rate between $0°$ and $30°$ and decreases slowly from angle $30°$ to $90°$, while the repeated folding curve remains nearly zero between $0°$ and $30°$ and increases at a higher rate after $30°$. The repeated folding curve reaches nearly the same maximum value as the virgin folding curve, which is preferred by the packaging industry which requires the repeated folded carton to preserve good material recovery and recycling qualities.

Figure 8.4 shows the motion-stiffness curves of virgin and repeated folding experiments, where the folding stiffness is the gradient of the motion-moment curve. Values at eight folding positions every $10°$ from $10°$ to $80°$ are selected to represent stiffness curves. Each error bar corresponds to the standard deviations of folding stiffness within $\pm 5°$ range at each folding position. The virgin folding stiffness increases sharply and is much higher than repeated folding stiffness between $0°$ and $30°$, then decreases slowly and keeps almost constant after $40°$. The repeated folding stiffness remains nearly zero before $30°$ due to the plastic deformation of creases and increases

Fig. 8.3 Folding moment comparison between virgin and repeated folding, solid and dashed lines represent experiment results, and dash-dot lines represent $\pm$ one standard deviation

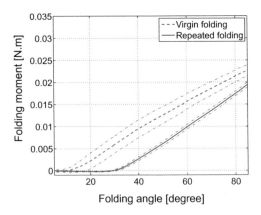

Fig. 8.4 Folding stiffness
comparison, solid and
dashed lines represent
experiment results, error bars
represent standard deviations
within ±5° range of eight
folding positions every 10°
from 10° to 80°

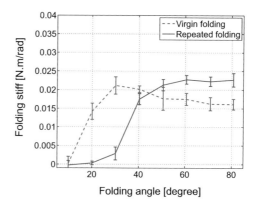

quickly between 30° to 40° and remains constant after 40°. Based on the comparisons
between virgin and repeated folding manipulations, repeated folding experiments are
more sustainable, and the following experimental results are selected from repeated
folding experiments of single creases.

8.2.2   Crease Stiffness with Various Crease Lengths

In this section, creases with various lengths were tested to explore their folding
behavior under the repeated folding condition. Creases with length $l_1 = 0.05$ m,
$l_2 = 0.075$ m, and $l_3 = 0.10$ m were folded with same angular velocity 1 rpm from
0° to 90°. For each length, five specimens were made and each was tested with
ten experiment runs, and experiment results from three to ten experiment runs were
selected to make comparisons. Folding stiffness results are compared in Fig. 8.5,
where stiffness with different crease lengths remain unchanged from 0° to 20°, and
then increase sharply from 20° to 40° and remain constant from 40° to 90°.

Fig. 8.5 Folding stiffness
comparison, seven folding
positions ranging from 20° to
80° are selected to represent
motion-stiffness curves

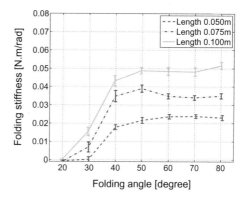

Table 8.1 Folding stiffness of single creases with different lengths

Crease	Crease length (m)	Average stiffness (N m rad^{-1})	Average stiffness per unit length (N rad^{-1})	Stiffness std. ($N = 40$)
1	0.050	0.023	0.47	1.3e–03
2	0.075	0.036	0.48	1.4e–03
3	0.100	0.048	0.48	2.0e–03

Further the average values of folding stiffness between 40° to 90° and stiffness per unit length are provided in Table 8.1, including standard deviations of stiffness values. From Table 8.1 we can see stiffness per unit length of three lengths are 0.47, 0.48 and 0.48 N rad^{-1}. MATLAB two-sample t test is applied to examine the correlations between each two results. The hypothesis test result $h = 0$ indicates a failure to reject the null hypothesis at 5% significance level, which suggests that the folding stiffness of crease is proportional to the length of crease.

8.2.3 Crease Stiffness with Various Folding Velocities

Following the experimental manipulations of creases with various lengths, folding characteristics of creases with various folding angular velocities were tested in this section. Crease length of specimens equaled to $l = 0.05$ m. The angular velocity of test 1 was set to be $\omega_1 = 1$ rpm, test 2 was set with angular velocity $\omega_2 = 2$ rpm and test 3 was $\omega_2 = 3.5$ rpm

The motion-moment characteristics of tests are presented in Fig. 8.6, which shows they overlap each other, and indicates the generated resisted moment of crease with various folding velocities remain almost the same. Further Fig. 8.6 demonstrates the motion-stiffness curves with various folding velocities increase sharply between 30°

Fig. 8.6 Folding stiffness comparison, dashed line, dash-dot line and solid line corresponds to experiment results, seven folding positions every 10° from 20° to 80° are selected to represent motion-stiffness curves

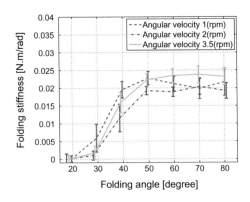

Table 8.2 Folding stiffness of creases with different folding velocities

Crease	Folding velocity (rpm)	Average stiffness (N m rad^{-1})	Average stiffness per unit length (N rad^{-1})	Stiffness std. ($N = 40$)
1	1.0	0.022	0.44	3.1e–03
2	2.0	0.021	0.42	1.7e–03
3	3.5	0.023	0.46	2.3e–03

and 40° and remain almost constant after 40°. Similarly average values of fold-ing stiffness between 40° and 90° and stiffness per unit length are provided in Table 8.2, where stiffness per unit length with three folding velocities are 0.44, 0.42 and 0.46 N rad^{-1} ($h = 1$ with two-sample t tests), which indicates we should take the folding angular velocity into consideration when modeling the single crease's folding characteristics.

8.3 A Framework for Reaction-Force Modeling of an Origami Mechanism

Section 8.2 presents a complete study of single creases of origami folds, which reveals an origami crease can be treated as a one-DOF flexible element with embedded tor-sional stiffness. This paves the way for the further force analysis of origami compliant structures using the proposed repelling-screw based approach.

In this section, the theoretical background of the repelling-screw based approach is introduced. When an origami structure is deformed, torques generated from folded creases result in an integrated reaction wrench to counter the external wrench applied at the origami structure. Thus an origami structure can be made equivalent to a redundantly actuated parallel mechanism whose revolute joints are supported by either active actuators or passive torsional springs. The process of calculating the reaction force from resistive torques of joints is named as the forward force analysis. Following the principle of equivalence, the forward force analysis developed for redundantly actuated parallel manipulators can be implemented in the reaction-force analysis of origami parallel mechanisms. Assuming an origami parallel mechanism has m limbs, and for the i-th limb it has $c_i (i < 6)$ creases as compliant joints, then the twist $\boldsymbol{T}_i$ of i-th limb has the form

$$\boldsymbol{T}_i = \Delta \mathbf{J}_s \delta \boldsymbol{\theta} = \Delta \begin{bmatrix} \boldsymbol{S}_{i,1} & \cdots & \boldsymbol{S}_{i,c_i} \end{bmatrix} \begin{bmatrix} \delta \theta_{i,1} \\ \vdots \\ \delta \theta_{i,c_i} \end{bmatrix} \tag{8.1}$$

where $\boldsymbol{J}_S$ is the Jacobian matrix of limb i and $\boldsymbol{S}_{i,j}$ ($j = 1, \ldots, c_i$) are the joint screws, they are 6×1 unit instantaneous twists written in Plücker ray coordinates whose first three elements represent rotational displacements and second three elements represent linear displacements. $\delta\theta_{i,j}$ is the rotational angle of the j-th crease. According to the reciprocal relationship [40], a *rank* $(6 - c_i)$ reciprocal screw system that is reciprocal to all joint screws of limb i can be obtained. These reciprocal screws are labeled as $\boldsymbol{w}^r_{i,k}$ ($k = 1, \ldots, 6 - c_i$) whose first three elements represent forces and last three elements represent moments, they are written in Plücker ray coordinates. The relationship between $\boldsymbol{w}^r_{i,k}$ and $\boldsymbol{T}_i$ can be written as

$$(\boldsymbol{w}^r_{i,k})^{\mathrm{T}}\boldsymbol{T}_i = 0 \tag{8.2}$$

Considering revolute joint j is fixed, a repelling screw $\boldsymbol{w}_{i,j}$ can be found that is reciprocal to the screws of all other joints and has a positive product with its own screw $\boldsymbol{S}_{i,j}$. The product of $\boldsymbol{w}_{i,j}$ with $\boldsymbol{T}_i$ can be written as

$$\boldsymbol{w}^{\mathrm{T}}_{i,j}\boldsymbol{T}_i = \boldsymbol{w}^{\mathrm{T}}_{i,j}\Delta\boldsymbol{S}_{i,j}\delta\theta_{i,j} \tag{8.3}$$

Also $\boldsymbol{w}_{i,j}$ is linearly independent of constraint wrenches $\boldsymbol{w}^r_{i,k}$. As a result, a total number of c_i repelling screws introduced by creases can be found and an equilibrium can be derived, that is

$$\begin{bmatrix} \boldsymbol{w}^{\mathrm{T}}_{i,1} \\ \vdots \\ \boldsymbol{w}^{\mathrm{T}}_{i,l_i} \end{bmatrix}\boldsymbol{T}_i = \begin{bmatrix} \boldsymbol{w}^{\mathrm{T}}_{1,1}\Delta\boldsymbol{S}_{1,1} & & \\ & \ddots & \\ & & \boldsymbol{w}^{\mathrm{T}}_{i,c_i}\Delta\boldsymbol{S}_{i,c_i} \end{bmatrix}\delta\boldsymbol{\theta} \tag{8.4}$$

where $\delta\boldsymbol{\theta} = \begin{bmatrix} \delta\theta_{i,1} & \cdots & \delta\theta_{i,c_i} \end{bmatrix}^{\mathrm{T}}$. According to Eq. (8.4), we can obtain the formula of $\delta\boldsymbol{\theta}$ as

$$\begin{aligned} \delta\boldsymbol{\theta} &= \begin{bmatrix} \boldsymbol{w}^{\mathrm{T}}_{1,1}\Delta\boldsymbol{S}_{1,1} & & \\ & \ddots & \\ & & \boldsymbol{w}^{\mathrm{T}}_{i,c_i}\Delta\boldsymbol{S}_{i,c_i} \end{bmatrix}^{-1}\begin{bmatrix} \boldsymbol{w}^{\mathrm{T}}_{i,1} \\ \vdots \\ \boldsymbol{w}^{\mathrm{T}}_{i,l_i} \end{bmatrix}\boldsymbol{T}_i \\ &= \begin{bmatrix} \dfrac{1}{\boldsymbol{w}^{\mathrm{T}}_{1,1}\Delta\boldsymbol{S}_{1,1}} & & \\ & \ddots & \\ & & \dfrac{1}{\boldsymbol{w}^{\mathrm{T}}_{i,c_i}\Delta\boldsymbol{S}_{i,c_i}} \end{bmatrix}\begin{bmatrix} \boldsymbol{w}^{\mathrm{T}}_{i,1} \\ \vdots \\ \boldsymbol{w}^{\mathrm{T}}_{i,l_i} \end{bmatrix}\boldsymbol{T}_i \end{aligned} \tag{8.5}$$

Assuming an external force is applied at the platform of origami mechanism, each limb generates a reaction force $\boldsymbol{w}_i$ to counter the external force. For the i-th limb, each revolute joint j generates a resistive torque $\tau_{i,j}$ and together integrates to the reaction force $\boldsymbol{w}_i$. In order to obtain the formula of $\boldsymbol{w}_i$, the instant power relationship $(\boldsymbol{w}_i)^{\mathrm{T}}\boldsymbol{T}_i = \boldsymbol{\tau}^{\mathrm{T}}\delta\boldsymbol{\theta}$ is utilized. Substituting Eq. (8.5) into the instant power relationship, we can obtain

$$(\boldsymbol{w}_i)^{\mathsf{T}} \boldsymbol{T}_i = \boldsymbol{\tau}^{\mathsf{T}} \begin{bmatrix} \dfrac{1}{\boldsymbol{w}_{i,1}^{\mathsf{T}} \Delta \boldsymbol{S}_{i,1}} & & \\ & \ddots & \\ & & \dfrac{1}{\boldsymbol{w}_{i,c_i}^{\mathsf{T}} \Delta \boldsymbol{S}_{i,c_i}} \end{bmatrix} \begin{bmatrix} \boldsymbol{w}_{i,1}^{\mathsf{T}} \\ \vdots \\ \boldsymbol{w}_{i,c_i}^{\mathsf{T}} \end{bmatrix} \boldsymbol{T}_i \tag{8.6}$$

where $\boldsymbol{\tau} = \begin{bmatrix} \tau_{i,1} & \cdots & \tau_{i,c_i} \end{bmatrix}^{\mathsf{T}}$. Eq. (8.6) leads to the formula of $\boldsymbol{w}_i$ as

$$\boldsymbol{w}_i = \begin{bmatrix} \boldsymbol{w}_{i,1} & \cdots & \boldsymbol{w}_{i,c_i} \end{bmatrix} \begin{bmatrix} \dfrac{1}{\boldsymbol{w}_{i,1}^{\mathsf{T}} \Delta \boldsymbol{S}_{i,1}} & & \\ & \ddots & \\ & & \dfrac{1}{\boldsymbol{w}_{i,c_i}^{\mathsf{T}} \Delta \boldsymbol{S}_{i,c_i}} \end{bmatrix} \begin{bmatrix} \tau_{i,1} \\ \vdots \\ \tau_{i,c_i} \end{bmatrix} \tag{8.7}$$

and it can further be simplified as

$$\boldsymbol{w}_i = \begin{bmatrix} \boldsymbol{w}_{i,1} & \cdots & \boldsymbol{w}_{i,c_i} \end{bmatrix} \begin{bmatrix} \dfrac{\tau_{i,1}}{\boldsymbol{w}_{i,1}^{\mathsf{T}} \Delta \boldsymbol{S}_{i,1}} \\ \vdots \\ \dfrac{\tau_{i,c_i}}{\boldsymbol{w}_{i,c_i}^{\mathsf{T}} \Delta \boldsymbol{S}_{i,c_i}} \end{bmatrix} = \sum_{j=1}^{c_i} \dfrac{\boldsymbol{w}_{i,j}}{\boldsymbol{w}_{i,j}^{\mathsf{T}} \Delta \boldsymbol{S}_{i,j}} \tau_{i,j} \tag{8.8}$$

Following the forward force analysis of one limb, the reaction force of the whole origami mechanism can be obtained to counter the external force applied, which is the sum of reaction forces generated by all m limbs as

$$\boldsymbol{w}_e = \sum_{i=1}^{m} \boldsymbol{w}_i = \sum_{i=1}^{m} \sum_{j=1}^{c_i} \dfrac{\boldsymbol{w}_{i,j}}{\boldsymbol{w}_{i,j}^{\mathsf{T}} \Delta \boldsymbol{S}_{i,j}} \tau_{i,j} \tag{8.9}$$

From Eq. (8.9) we can see $\boldsymbol{w}_e$ is a function of the twists $\boldsymbol{S}_{i,j}$, repelling screws $\boldsymbol{w}_{i,j}$ and resistive torques $\tau_{i,j}$. $\boldsymbol{S}_{i,j}$ and $\boldsymbol{w}_{i,j}$ are purely based on the geometry and position of folded origami structure, while $\tau_{i,j}$ depends on the elastic properties of crease j of limb i. Without lose of generality, a layer-type crease [33, 34] is considered here, whose corresponding $\tau_{i,j}$ can be written as

$$\tau_{i,j} = k_0 l_{i,j} (\theta_{i,j} - \theta_{i,j}(0)) \tag{8.10}$$

where k_0 is the stiffness coefficient per unit length, $l_{i,j}$ is the length of crease j, $\theta_{i,j}(0)$ and $\theta_{i,j}$ are angles between two panels of crease j at the initial and current positions. k_0 is assumed to be constant here. In case where nonlinear torsional stiffness needs to be addressed, we can set k_0 as an angle dependent variable. The resistive torques generated by other types of creases can be developed in a similar manner but with different stiffness coefficients. As a result, the elastic performance of single creases are integrated into the reaction force calculation via Eq. (8.10). Following the analytical force modelling of origami structures, examples are given in the following sections to demonstrate the utility of the developed modelling approach. Particularly the widely used waterbomb-origami linkage [15, 27, 37] is selected to demonstrate the detailed algebraic-calculation process, and the calculation result will be further used in the modelling of a waterbomb-origami integrated parallel mechanism [41].

8.4 Reaction Force of an Origami Linkage Based on Its Repelling Screws

In this section, a detailed algebraic calculation is provided to model the reaction force of the widely used origami-waterbomb linkage. Repelling screws of the origami linkage are developed first, followed by the reaction-force calculation based on the obtained repelling screws.

8.4.1 Repelling Screws of the Origami-Waterbomb Linkage

Following the mechanism-equivalent approach, the origami-waterbomb fold under an external force and a schematic diagram of its equivalent mechanism are shown in Fig. 8.7. From it we can see the waterbomb-origami fold forms a spherical linkage consisting of six revolute joints $S_i(i = 1, \ldots, 6)$. The lower panel formed by S_1 and S_4 is connected to the base via a revolute joint B, and the upper panel formed by S_3 and S_6 has a revolute joint P. Both lower and upper panels move symmetrically to a plane H formed by revolute joints S_2 and S_5. If an external force is applied at the upper panel through revolute joint P, it is countered by reaction forces of two kinematic chains. Kinematic chain E_1AF_1 contains revolute joints $S_i(i = 1, \ldots, 3)$ and kinematic chain E_2AF_2 contains revolute joints $S_i(i = 4, \ldots, 6)$.

Fig. 8.7 The equivalent mechanism of an origami waterbomb base

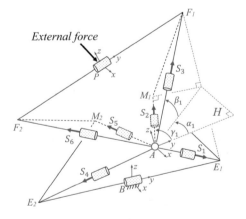

The revolute joints contained in chain E_1AF_1 are described in the coordinate frame $\{A, x, y, z\}$, where the x axis is normal to the plane $E_1F_1F_2$, y axis is parallel to $E_1F_1F_2$, and z axis is determined according to the right-hand-rule of Cartesian coordinate. The screw of each revolute joint in chain one can be written as

$$
\begin{aligned}
S_1 &= \begin{bmatrix} -\cos\beta_1\sin\alpha_1 & \cos\beta_1\cos\alpha_1 & -\sin\beta_1 & 0 & 0 & 0 \end{bmatrix}^{\mathrm{T}} \\
S_2 &= \begin{bmatrix} -\sin\gamma_1 & \cos\gamma_1 & 0 & 0 & 0 & 0 \end{bmatrix}^{\mathrm{T}} \\
S_3 &= \begin{bmatrix} -\cos\beta_1\sin\alpha_1 & \cos\beta_1\cos\alpha_1 & \sin\beta_1 & 0 & 0 & 0 \end{bmatrix}^{\mathrm{T}}
\end{aligned}
\tag{8.11}
$$

from which we can see they present three axes of pure rotations. α_1 presents the angle between the projection line of screw S_1 and the y axis on the $x - y$ plane at the coordinate frame $\{A, x, y, z\}$. β_1 is the projection angle of screw S_1 on the symmetric plane H, γ_1 is the angle between S_2 and y axis. Three constraint wrenches can then be identified according to their reciprocal relationship with screws in Eq. (8.11). Their simplest forms are adopted here as

$$
\begin{aligned}
w_1^r &= \begin{bmatrix} 1 & 0 & 0 & 0 & 0 & 0 \end{bmatrix}^{\mathrm{T}} \\
w_2^r &= \begin{bmatrix} 0 & 1 & 0 & 0 & 0 & 0 \end{bmatrix}^{\mathrm{T}} \\
w_3^r &= \begin{bmatrix} 0 & 0 & 1 & 0 & 0 & 0 \end{bmatrix}^{\mathrm{T}}
\end{aligned}
\tag{8.12}
$$

where w_i^r is written in Plücker ray coordinate whose first part is a 3×1 force vector and second part is a 3×1 moment vector. It can be seen these three constraint wrenches given in Eq. (8.12) are pure forces passing through the origin of coordinate frame $\{A, x, y, z\}$. When one revolute joint is fixed, the corresponding repelling screw is introduced that is reciprocal to the other two twists and independent of the constraint wrenches. By fixing revolute joint S_1, wrench w_1 carried by the repelling screw can be written as

$$
w_1 = \begin{bmatrix} 0 & 0 & 0 & x & y & z \end{bmatrix}^{\mathrm{T}}
\tag{8.13}
$$

where x, y and z satisfy

$$
\begin{aligned}
(-\sin\gamma_1)x + (\cos\gamma_1)y &= 0 \\
(-\sin\beta_1\sin\alpha_1)x + (\cos\beta_1\cos\alpha_1)y + (\sin\beta_1)z &= 0
\end{aligned}
\tag{8.14}
$$

A solution of w_1 has the form

$$
w_1 = \begin{bmatrix} 0 & 0 & 0 & \sin\beta_1\cos\gamma_1 & \sin\beta_1\sin\gamma_1 & \cos\beta_1\sin(\alpha_1 - \gamma_1) \end{bmatrix}^{\mathrm{T}}
\tag{8.15}
$$

w_2 and w_3 can be obtained similarly by fixing S_2 and S_3 and calculated with their corresponding repelling screws, the following can be given

$$
\begin{aligned}
w_2 &= \begin{bmatrix} 0 & 0 & 0 & -\cos\beta_1\cos\alpha_1 & -\cos\beta_1\sin\alpha_1 & 0 \end{bmatrix}^{\mathrm{T}} \\
w_3 &= \begin{bmatrix} 0 & 0 & 0 & \sin\beta_1\cos\gamma_1 & \sin\beta_1\sin\gamma_1 & -\cos\beta_1\sin(\alpha_1 - \gamma_1) \end{bmatrix}^{\mathrm{T}}
\end{aligned}
\tag{8.16}
$$

From which we can see the repelling screws carried three pure torque constraints passing through origin A. After developing repelling screws of the first kinematic chain, we can analyze the second kinematic chain of the waterbomb linkage in the same way. The second kinematic chain E_2AF_2 contains three revolute joints $S_i (i = 4, 5, 6)$, whose screws have the forms as

$$
\begin{aligned}
S_4 &= \begin{bmatrix} -\cos \beta_2 \sin(\pi - \alpha_2) & \cos \beta_2 \cos(\pi - \alpha_2) & -\sin \beta_2 & 0 & 0 & 0 \end{bmatrix}^{\mathrm{T}} \\
S_5 &= \begin{bmatrix} -\sin(\pi - \gamma_2) & \cos(\pi - \gamma_2) & 0 & 0 & 0 & 0 \end{bmatrix}^{\mathrm{T}} \\
S_6 &= \begin{bmatrix} -\cos \beta_2 \sin(\pi - \alpha_2) & \cos \beta_2 \cos(\pi - \alpha_2) & \sin \beta_2 & 0 & 0 & 0 \end{bmatrix}^{\mathrm{T}}
\end{aligned}
\tag{8.17}
$$

where α_2 is the angle between the projection line of screw S_4 and the $-y$ axis on the $x - y$ plane, β_2 is the projection angle of screw S_4 on the symmetric plane H, and γ_2 is the angle between S_5 and $-y$ axis. It is easy to identify that the constraint wrenches in Eq. (8.12) are also reciprocal to the screws in Eq. (8.17), thus they are the common constraint wrenches of waterbomb linkage A. Subsequently, the introduced repelling screws can be developed similarly by fixing their corresponding revolute joints, which can be written as

$$
\begin{aligned}
w_4 &= [0 \quad 0 \quad 0 \quad \cdots \\
&\quad -\sin \beta_2 \cos(\pi - \gamma_2) \quad -\sin \beta_2 \sin(\pi - \gamma_2) \quad \cos \beta_2 \sin(\alpha_2 - \gamma_2)]^{\mathrm{T}} \\
w_5 &= [0 \quad 0 \quad 0 \quad \cos \beta_2 \cos(\pi - \alpha_1) \quad \cos \beta_2 \sin(\pi - \alpha_2) \quad 0]^{\mathrm{T}} \\
w_6 &= [0 \quad 0 \quad 0 \quad \cdots \\
&\quad -\sin \beta_2 \cos(\pi - \gamma_2) \quad -\sin \beta_2 \sin(\pi - \gamma_2) \quad -\cos \beta_2 \sin(\alpha_2 - \gamma_2)]^{\mathrm{T}}
\end{aligned}
\tag{8.18}
$$

The repelling screws in Eq. (8.18) represent other three constraint torques.

8.4.2 Reaction Force of the Waterbomb-Origami Linkage

The obtained repelling screws $w_i (i = 1, \ldots, 6)$ can be used to develop the reaction force of the waterbomb linkage. The reaction force of kinematic chain E_1AF_1 can be expressed as

$$
w_{L,1} = \begin{bmatrix} w_1 & w_2 & w_3 \end{bmatrix} \begin{bmatrix} \frac{1}{w_1^T \Delta S_1} & 0 & 0 \\ 0 & \frac{1}{w_2^T \Delta S_2} & 0 \\ 0 & 0 & \frac{1}{w_3^T \Delta S_3} \end{bmatrix} \begin{bmatrix} \tau_1 \\ \tau_2 \\ \tau_3 \end{bmatrix}
\tag{8.19}
$$

where τ_i is the resistive torque generated by crease $i (i = 1, 2, 3)$. The obtained repelling screws $w_i (i = 4, 5, 6)$ can be used to develop the reaction force of kinematic chain E_2AF_2 as

$$\boldsymbol{w}_{L,2} = \begin{bmatrix} \boldsymbol{w}_4 & \boldsymbol{w}_5 & \boldsymbol{w}_6 \end{bmatrix} \begin{bmatrix} \frac{1}{\boldsymbol{w}_4^T \Delta S_4} & 0 & 0 \\ 0 & \frac{1}{\boldsymbol{w}_5^T \Delta S_5} & 0 \\ 0 & 0 & \frac{1}{\boldsymbol{w}_6^T \Delta S_6} \end{bmatrix} \begin{bmatrix} \tau_4 \\ \tau_5 \\ \tau_6 \end{bmatrix} \tag{8.20}$$

As the resistive forces generated by both kinematic chains are developed, they together generate the reaction force of this waterbomb linkage A formed by these two kinematic chains

$$\boldsymbol{w}_A = \boldsymbol{w}_{L,1} + \boldsymbol{w}_{L,2} = \sum_{i=1}^{6} \frac{\boldsymbol{w}_i}{\boldsymbol{w}_i^T \Delta S_i} \tau_i \tag{8.21}$$

From Eqs. (8.19) and (8.20) we can see the reaction force $\boldsymbol{w}_A$ is a pure torque, meaning this spatial linkage is equivalent to a 3 DOF spherical joint. It implies that an origami waterbomb base can be treated as a spherical joint with resistive torques in all three rotation directions.

8.4.3 Geometric Relations

As in Eq. (8.21), reaction force $\boldsymbol{w}_A$ is determined by screws S_i, repelling screws $\boldsymbol{w}_i$ and resistive torques from creases $\tau_i (i = 1, \ldots, 6)$. The analytical models of S_i and $\boldsymbol{w}_i$ can be calculated using position angles α_j, β_j and $\gamma_j (j = 1, 2)$, whose geometrical relationships are not straightforward and need a further calculation. As in Fig. 8.8, by introducing the symmetric plane H, their geometrical relationships can be established as

$$\begin{aligned} \sin \beta_j &= \tfrac{h_j}{d} \\ \cos \beta_j \cos(\gamma_j - \alpha_j) &= \tfrac{a}{d} \\ \cos \beta_j \sin \alpha_j &= \tfrac{1}{d}\sqrt{b^2 - \tfrac{1}{4}(h_1 + h_2)^2} \end{aligned} \tag{8.22}$$

where h_j is half the length of line segment $E_j F_j$ and equals to the projection distance between vertex E_j and plane H. It will be shown this plane H is essential in the further modelling of an origami-waterbomb based parallel platform. a is the length of AM_1, b is the length of $M_1 F_1$ and d is the length of AF_1. Here we assume h_j is known and all positioning angles can be determined according to Eq. (8.22). Then we turn to the resistive torque τ_i generated by crease i. According to Eq. (8.10), τ_i has the form

$$\tau_i = k_0 l_i (\theta_{s,i} - \theta_{s,i}(0)) \tag{8.23}$$

in which k_0 and l_i are decided by the material properties and length of crease i and are pre-determined [34]. $\theta_{s,i}(0)$ is the initial angle between two panels of crease i, $\theta_{s,i}$ is the angle at the current stage and determined by the position of waterbomb

Fig. 8.8 Geometrical
relationships of the
equivalent spherical joint A

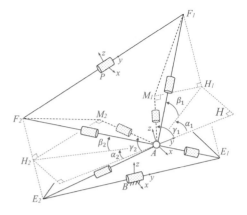

linkage. Take crease 1 for example, its two panels are labeled as AB_1E_1 and AE_1M_1. AB_1E_1 is formed by screws S_1 and S_4, AE_1M_1 is formed by screws S_1 and S_2. The angle between two panels is also equal to the angle between normal vectors of them, which have the forms as $S_4 \times S_1$ and $S_1 \times S_2$. Thus the angle of crease 1 can be calculated as

$$\theta_{s,1} = atan2\{\|(S_4 \times S_1) \times (S_1 \times S_2)\|, (S_4 \times S_1)^T (S_1 \times S_2)\} \tag{8.24}$$

which is always the smallest angle between two vectors $S_4 \times S_1$ and $S_1 \times S_2$. Rotation angles of other creases can be calculated in a similar manner. Consequently, the reaction force w_A of waterbomb linkage A is fully determined.

8.5 Reaction-Force Modeling of a Waterbomb Parallel Mechanism

It is noticed that most developed origami structures are of integrated waterbomb-origami patterns [14, 15, 27], thus it is necessary to conduct a further force analysis of origami mechanisms based on the integration of waterbomb linkages.

8.5.1 Kinematic Description of the Waterbomb Parallel Mechanism

The selected origami structure is based on the parallel integration of three origami waterbomb panels. Similar origami structures that are formed by waterbomb bases can also be found in [14, 15]. The kinematic diagram of the equivalent mechanism is shown in Fig. 8.9, where the mechanism consists of a base and a moving platform

Fig. 8.9 Schematic diagram
of the waterbomb-base
integrated parallel
mechanism

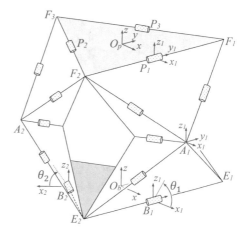

and they are connected by three limbs in parallel. Each limb is labeled as $B_i A_i P_i (i = 1, 2, 3)$, B_i and P_i denote single creases which are treated as revolute joints. A_i is formed with six creases which is the same waterbomb-origami linkage analyzed in Sect. 8.4. Since a waterbomb linkage A_i can be taken as a spherical joint, each limb is then equivalent to an RSR serial chain and the whole mechanism to a 3-RSR parallel platform [42]. Both the base and platform are designed to have the same equilateral-triangle shape with the radius of in circle equal to r. The geometrical size of the waterbomb linkage follows the definition used in Sect. 8.4.3.

The moving platform is revealed to have three degrees of freedom, including one translation and two rotations [27]. As a result, the motion of the platform can be fully determined by three input rotation angles $\theta_i (i = 1, 2, 3)$ of revolute joint B_i. Due to the symmetrical structure of waterbomb linkages, the base and moving platform are symmetric to the plane H formed by vertices A_i. Thus the symmetric plane H can be used to describe the motion of the platform, its detailed formulation is given in Appendix D. The formulation of H is further used to describe the motion of this origami mechanism in Appendix E. When an external load is applied at the upper base, the origami structure will deform following the allowed degrees of freedom.

8.5.2   Repelling Screws of the Waterbomb Parallel Mechanism

Similar to the repelling screw construction of waterbomb linkage A conducted in Sect. 8.4.1, screws of each limb $B_i A_i P_i$ are developed first. Coordinate frame $\{A_i, x_i, y_i, z_i\}$ is used here in order to simplify the screw representation of each joint in the waterbomb linkage. As a result, the screws of spherical joint A_i can be written as

$$S_{Ai,x} = \begin{bmatrix} 1 & 0 & 0 & 0 & 0 & 0 \end{bmatrix}^{\mathrm{T}}$$
$$S_{Ai,y} = \begin{bmatrix} 0 & 1 & 0 & 0 & 0 & 0 \end{bmatrix}^{\mathrm{T}} \tag{8.25}$$
$$S_{Ai,z} = \begin{bmatrix} 0 & 0 & 1 & 0 & 0 & 0 \end{bmatrix}^{\mathrm{T}}$$

which are along the axes of $\{A_i, x_i, y_i, z_i\}$. In terms of revolute joints B_i and P_i, they are symmetric about virtual plane H. If revolute joint B_i is assumed to have a direction vector (l, m, n) and position vector (r_x, r_y, r_z), screws of revolute joints B_i and P_i can be written as

$$S_{Bi} = \begin{bmatrix} l & m & n & P & Q & R \end{bmatrix}^{\mathrm{T}}$$
$$S_{Pi} = \begin{bmatrix} l & m & -n & -P & -Q & R \end{bmatrix}^{\mathrm{T}} \tag{8.26}$$

where $P = nr_y - mr_z$, $Q = lr_z - nr_x$ and $R = mr_x - lr_y$. The screws of the equivalent spherical joint A_i and revolute joints B_i and P_i span a $rank$ 5 screw system, which has one reciprocal constraint wrench. The constraint wrench that is reciprocal to all screws in Eqs. (8.25) and (8.26) can be calculated as

$$w_1^r = \begin{bmatrix} -Q & P & 0 & 0 & 0 & 0 \end{bmatrix}^{\mathrm{T}} \tag{8.27}$$

The next step is to calculate the repelling screws that are introduced by locking axes of spherical joint A_i as well as revolute joints B_i and P_i. Fixing the corresponding screw, the resulted repelling screws can be calculated that is reciprocal to the rest screws while linearly independent of w_c. The repelling screws of equivalent spherical joint A_i are

$$w_{Ai,x} = \begin{bmatrix} 0 & 0 & -l & R & 0 & 0 \end{bmatrix}^{\mathrm{T}}$$
$$w_{Ai,y} = \begin{bmatrix} 0 & 0 & -m & 0 & R & 0 \end{bmatrix}^{\mathrm{T}} \tag{8.28}$$
$$w_{Ai,z} = \begin{bmatrix} -\dfrac{P}{\sqrt{P^2+Q^2}}n & -\dfrac{Q}{\sqrt{P^2+Q^2}}n & 0 & 0 & 0 & \sqrt{P^2+Q^2} \end{bmatrix}^{\mathrm{T}}$$

Similarly, the repelling screws of revolute joints B_i and P_i can be written as

$$w_{Bi} = \begin{bmatrix} PR & QR & P^2+Q^2 & 0 & 0 & 0 \end{bmatrix}^{\mathrm{T}}$$
$$w_{Pi} = \begin{bmatrix} -PR & -QR & P^2+Q^2 & 0 & 0 & 0 \end{bmatrix}^{\mathrm{T}} \tag{8.29}$$

It is noticed that only simple representations of screws S_{Bi} and S_{Pi} are given in Eq. (8.26). In order to obtain their detailed analytical values, their screw-vector representations in coordinate frame $\{A_i, x_i, y_i, z_i\}$ are needed. Figure 8.10 shows the geometrical relationships between different coordinate frames. Here a adjoint matrix $\mathbf{Ad}_{bi}$ [43] is used that transforms S_{B1} from coordinate frame $\{B_i, x_i, y_i, z_i\}$ to $\{A_i, x_i, y_i, z_i\}$, which can be written as

$$S_{Bi} = \Delta(\mathbf{Ad}_{bi})^{-1}\Delta S'_{Bi} \tag{8.30}$$

Fig. 8.10 Geometrical relationships of the Limb $B_1 A_1 P_1$

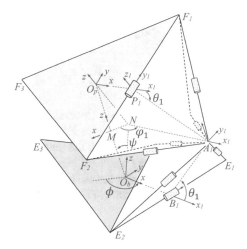

where $S'_{Bi} = \begin{bmatrix} 0 & 0 & 0 & 0 & 1 & 0 \end{bmatrix}^T$ and it is written in local coordinate frame and Δ is the operator [44] that exchanges the first and second parts of a screw. Adjoint matrix $\mathbf{Ad}_{bi}$ has the form

$$\mathbf{Ad}_{bi} = \begin{bmatrix} \mathbf{R}_{bi} & \mathbf{0} \\ \mathbf{P}_{bi}\mathbf{R}_{bi} & \mathbf{R}_{bi} \end{bmatrix} \tag{8.31}$$

$\mathbf{R}_{bi}$ is the coordinate rotation matrix, and $\mathbf{P}_{bi}$ is the anti-symmetric matrix representation of coordinate translation vector $\mathbf{p}_{bi}$. For joint $B_i (i = 1, 2, 3)$,

$$\mathbf{R}_{bi} = \mathbf{R}_z \left(\frac{2\pi}{3}(i - 1) + \phi \right) \mathbf{R}_y(\psi)\mathbf{R}_z(\varphi_i)$$
$$\mathbf{p}_{bi} = \begin{bmatrix} b\cos\theta_i & 0 & b\sin\theta_i \end{bmatrix}^T \tag{8.32}$$
$$\mathbf{p}_{bi} = \begin{bmatrix} b\cos\theta_i & 0 & b\sin\theta_i \end{bmatrix}^T$$

where ϕ and ψ are the two Euler angles used to describe the orientation of the moving platform, φ_i is the angle between $\overrightarrow{NM}$ and $\overrightarrow{NA_i}$. The formulations of these angles are given in Appendix E. S_{Pi} can be developed in a similar manner, it can also be calculated simply according to its symmetrical relationship with S_{Bi} in Eq. (8.26).

8.5.3 Reaction Force of the Waterbomb Parallel Mechanism

Based on the developed repelling screws, the resulted force of one limb $B_i A_i P_i$ can be calculated as

$$
\boldsymbol{w}_{L,i} = \left[\frac{\boldsymbol{w}_{Ai,x}}{\boldsymbol{w}_{Ai,x}^T \Delta \boldsymbol{S}_{Ai,x}} \quad \frac{\boldsymbol{w}_{Ai,y}}{\boldsymbol{w}_{Ai,y}^T \Delta \boldsymbol{S}_{Ai,y}} \quad \frac{\boldsymbol{w}_{Ai,z}}{\boldsymbol{w}_{Ai,z}^T \Delta \boldsymbol{S}_{Ai,z}} \right] \begin{bmatrix} \tau_{Ai,x} \\ \tau_{Ai,y} \\ \tau_{Ai,z} \end{bmatrix}
$$

$$
+ \left[\frac{\boldsymbol{w}_{Bi}}{\boldsymbol{w}_{Bi}^T \Delta \boldsymbol{S}_{Bi}} \quad \frac{\boldsymbol{w}_{Pi}}{\boldsymbol{w}_{Pi}^T \Delta \boldsymbol{S}_{Pi}} \right] \begin{bmatrix} \tau_{Bi} \\ \tau_{Pi} \end{bmatrix}
\tag{8.33}
$$

where $\tau_{Ai,x}$, $\tau_{Ai,y}$ and $\tau_{Ai,z}$ are resistive torques of equivalent spherical joint A_i, they can be obtained according to Eq. (8.21). τ_{B_i}, τ_{P_i} are resistive torques of revolute joints Bi and Pi. Following the reaction force calculation of one limb $B_i A_i P_i$, the reaction force of the whole origami mechanism can be established, which is the sum of three limbs' reaction forces. In the global coordinate frame $\{O_b, x, y, z\}$, the force equilibrium of the platform can be written as

$$
\boldsymbol{w}_e = \sum_{i=1}^{3} \mathbf{Ad}_{ai} \boldsymbol{w}_{L,i}
\tag{8.34}
$$

where $\boldsymbol{w}_{L,i}$ is the reaction force of limb $B_i A_i P_i$ described in the local coordinate frame $\{A_i, x_i, y_i, z_i\}$ and $\mathbf{Ad}_{ai}$ is the adjoint matrix from $\{O_b, x, y, z\}$ to $\{A_i, x_i, y_i, z_i\}$. $\mathbf{Ad}_{ai}$ has the same form as $\mathbf{Ad}_{bi}$ shown in Eq. (8.31), whose rotation matrix $\mathbf{R}_{ai}$ and translation vector $\mathbf{p}_{ai}$ can be written as

$$
\begin{aligned}
\mathbf{R}_{ai} &= \mathbf{R}_z(\phi)\mathbf{R}_y(\psi)\mathbf{R}_z(\varphi_i) \\
\mathbf{p}_{ai} &= \overrightarrow{O_b B_i} + \overrightarrow{B_i A_i}
\end{aligned}
\tag{8.35}
$$

8.6 Reaction-Force Evaluation Using Finite Element Simulations

Section 8.3 gives the framework of conducting forward-force analysis of compliant parallel mechanisms using repelling screws, making it possible to establish force-equilibrium when a compliant mechanism is under large deformation. To implement the developed approach, detailed calculation examples are given in Sects. 8.4 and 8.5, which successfully describes the reaction force of a compliant mechanism using the resistive torques of revolute joints when it is deformed. To validate the proposed repelling-screw based approach, in this section, we use finite element simulations to evaluate reaction forces of the waterbomb parallel mechanism in all degrees of freedom, including both the linear-motion mode and rotational-motion mode. The linear-motion mode happens when the origami mechanism presents purely up-down linear motion under a vertical external load, while the rotational-motion mode happens when the origami mechanism rotates around an arbitrary axis that is parallel to the base plane under external loads.

8.6.1   Reaction-Force Evaluation of the Linear-Motion Mode

The finite element simulation is conducted in the ANSYS transient-structural-analysis environment. A set of design parameters of the origami mechanism is provided in Table 8.3, including both the geometrical parameters of the whole structure and the stiffness properties of single creases. Specifically, according to Eq. (8.23), stiffness coefficient k_0 of single crease is assumed to be constant with loss of generality; crease lengths of crease 1, 3, 4, 6 in all waterbomb linkages are assumed to be equal and they are symbolized as l_l; crease 2 and 5 are assumed to have same length l_s. Stiffnesses of creases include B_i and P_i are omitted to simplify the comparison procedure. In addition, the initial balanced position of origami mechanism is determined when the rotation angles of revolute joint B_i ($i = 1, 2, 3$) are set to be $\theta_i(0) = 60°$. This configuration is used to calculate the initial rotation angle $\theta_s(0)$ of each crease.

Regarding the finite element simulation, geometrical parameters and stiffness constants of creases in waterbomb linkage are set according to the design parameters listed in Table 8.3. In addition, three extra prismatic joints between $E_i F_i$ ($i = 1, 2, 3$) are added in the origami mechanism for the purpose of adding external loads. When displacement load l_i is applied at each prismatic joint, the reaction force of the moving platform can be estimated by integrating their corresponding axial forces f_i.

The reaction force of linear-motion mode of origami mechanism is simulated by adding the same displacement loads at three prismatic joints, here they are set to be $l_i = 18$ mm ($i = 1, 2, 3$). The deformation diagram is presented in Fig. 8.11. f_1 and f_2 in Fig. 8.11 are the resulted axial forces of prismatic joints $E_1 F_1$ and $E_2 F_2$. Together with f_3, all three axial forces counter the reaction force w_e caused by resistive torques of creases in waterbomb-origami linkages.

The analytical solution of reaction force w_e can be developed by submitting both design parameters and displacement loads into related formulations. Due to sym-

Table 8.3 Design parameters of the origami mechanism

Platform	a (mm)	25	b (mm)	25	r (mm)	25
Crease	k_0 (N/rad)	0.45	l_l (mm)	35.4	l_s (mm)	25
Initial position	$\theta_1(0)$ (°)	60	$\theta_2(0)$ (°)	60	$\theta_3(0)$ (°)	60

Fig. 8.11 Linear motion of the origami mechanism

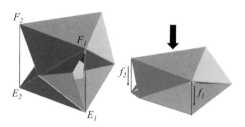

metry, the symbolic formula of the reaction force of one waterbomb linkage $\boldsymbol{w}_{Ai}$ in Eq. (8.21) can be yielded as

$$\boldsymbol{w}_{Ai} = \begin{bmatrix} 0 & 0 & 0 & 0 & \frac{2\sin\gamma_1\cdot\tau_L + 2\cos\beta_1\sin\alpha_1\cdot\tau_S}{\cos\beta_1\sin(\gamma_1-\alpha_1)} & 0 \end{bmatrix}^{\mathrm{T}} \tag{8.36}$$

Substituting Eq. (8.36) into Eq. (8.34), we are able to obtain the reaction force $\boldsymbol{w}_e$ which has the form

$$\boldsymbol{w}_e = \begin{bmatrix} 0 & 0 & f_e & 0 & 0 & 0 \end{bmatrix}^{\mathrm{T}} \tag{8.37}$$

where f_e is the magnitude of reaction force that is along the vertical axis of moving platform, it can be written as

$$f_e = \frac{3}{b\cos\theta_1}\frac{(2\sin\gamma_1\cdot\tau_L + 2\cos\beta_1\sin\alpha_1\cdot\tau_S)}{\cos\beta_1\sin(\gamma_1-\alpha_1)} \tag{8.38}$$

Subsequently FEA simulation is conducted by adding the displacement loads into the origami mechanism and measure the resulted axial force in each prismatic joint. In order to obtain a comprehensive evaluation result, both the rotation angles of creases and reaction force of the functional platform were compared and they are presented in Figs. 8.12 and 8.13. Since all waterbomb linkages have the same deformation, only crease 1 and crease 2 of waterbomb linkage A_1 are selected, and their rotation angles with respect to θ_1 of revolute joint B_1 are compared in Fig. 8.12. The comparison result suggests the FEA simulation results match analytical models quite well. Also, crease 1 and 2 have different rotation angles with respect to θ_1 though they are almost linear.

The integrated reaction force is further compared in Fig. 8.13, which shows a significant overlap between FEA simulation and analytical values. In addition, it is revealed the reaction force increases sharply at the early beginning and increases slowly with the increase of displacement loads.

Fig. 8.12 Crease rotation angles θ_s with respect to θ_1

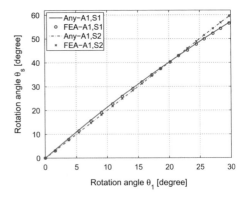

Fig. 8.13 Reaction force
with respect to θ_1

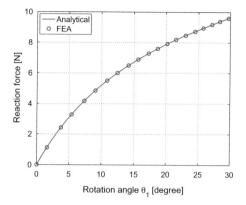

8.6.2 Reaction-Force Evaluation of the Rotational-Motion Mode

The reaction force of the origami mechanism in rotational-motion mode is evaluated in this section. The rotational motion of platform is illustrated in Fig. 8.14, where three prismatic joints are added with different displacement loads, as $l_2 = 18$ mm and $l_1 = l_3 = 0$ mm. f_1 and f_2 are the resulted axial forces of prismatic joints E_1F_1 and E_2F_2, ideally $f_1 = f_3 = 0$, and f_2 is the main force to counter reaction force $\boldsymbol{w}_e$ of the whole platform.

Unlike the linear-motion mode, the analytical model of $\boldsymbol{w}_e$ is rather complex and its symbolic formula is omitted here. FEA simulation is conducted by adding the displacement loads into the origami mechanism. Similarly, both the rotation angles of creases and the reaction force of functional platform were compared and they are presented in Figs. 8.15 and 8.16. Rotational angles of creases are compared in Fig. 8.15. Due to the unsymmetric displacement loads, waterbomb linkage A_1 and A_2 have symmetric deformation while no deformation occurs in waterbomb linkage $A_3(L_1 = L_3 = 0$ mm). As a result, creases 1, 2, 4, 5 in waterbomb linkages A_1 are selected to compare their rotation angles. The comparison result shows a good agreement between FEA simulation results and analytical models. It is noticed that not all creases have positively increased rotation angles. For example, the angle

Fig. 8.14 Rotational motion
of the origami mechanism

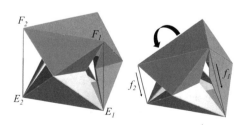

Fig. 8.15 Crease rotation angles θ_s with respect to θ_1

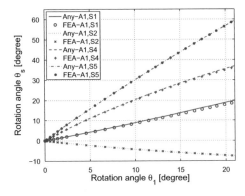

Fig. 8.16 Reaction force with respect to θ_1

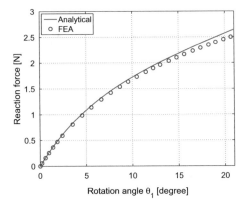

between two panels of crease 2 is decreasing with the increase of θ_1, which suggests it generate an opposite resistive torque compared to other creases.

Further, the integrated reaction force is compared in Fig. 8.16, where the analytical curve also matches the FEA simulation result. Though discrepancies exist at the end of rotation motion, they remain as small as 4%. In summary, the comparison results of both linear-motion and rotation-motion of the presented origami mechanism reveal the analytical models can match finite-element-simulations quite accurately, indicating the validity of the screw-theory based modelling approach developed in Sect. 8.3.

8.6.3 Applications of the Reaction-Force Evaluation

Though the main purpose of the reaction-force evaluation is to validate the proposed analytical modelling approach, there are some potential applications of the revealed motion-force characteristics. For example, the force evaluation results can be directly implemented into the design of origami-enabled continuum structures in [15–17]. These continuum manipulators followed similar module-based design principals, which integrate multiple layers of origami-module and actuate the whole structure through tendons or shape-memory-alloys. These actuators can be replaced by the three prismatic joints $E_i F_i$ in the presented simulation model if other internal forces such as friction are ignored. Linear-motion mode corresponds to the scenario when all three tendons are pulled with the same distance; rotational-motion mode corresponds to the scenario when tendons are pulled with different distances. Based on the developed analytical models, it is possible to estimate reaction force in each actuator and thus the reaction force of the whole platform, which are useful in the payload design as well as further development of advanced control strategy. In addition, linear-motion mode provided in Sect. 8.6.1 can be applied to many other applications, such as energy absorbers in [9] and the deformable wheel robot in [14]. The provided symbolic formulation of reaction force makes it possible to conduct optimization design such as the dimensional optimization of the thin-wall structure, as well as improved selection of shape-memory-alloy actuators in origami-wheel robots.

8.7   Conclusions

This chapter provided an approach for modelling the reaction force of origami structures when they are deformed. The approach makes a step forward from the mechanism-equivalent approach by treating origami structures as redundantly actuated parallel platforms and introduced the repelling screws to conduct force modelling of origami structures for the first time.

The theoretical framework of the force modelling approach is presented to determine the repelling screws introduced by creases in folding. These were then used to integrate into the reaction force of an origami, making it possible to include the folding behaviours of creases in a systematic way. The paper hence selected representative origami patterns to implement the modelling approach, including the widely used waterbomb base and its integrated parallel mechanism. The equivalent kinematics were then given, followed by reaction-force modelings using the repelling screws of folded creases.

Finite element simulations were conducted to validate the reaction-force. Both linear-motion and rotational-motion modes of the parallel origami mechanism were used to evaluate their reaction forces. The evaluation results in both scenarios demonstrated a good agreement between analytical models and simulation results. The presented approach combines the study of origami kinematics and that of elastic-performance of folded creases in the same theoretical framework, making it applicable to other origami mechanisms, either passive origami structures or active origami robots, with various origami patterns and elastic properties of creases.

References

1. Kanade, T.: A theory of origami world. Artif. Intell. **13**(3), 279–311 (1980)
2. Lang, R.J., Hull, T.C.: Origami design secrets: mathematical methods for an ancient art. Math. Intell. **27**(2), 92–95 (2005)
3. Dai, J., Jones, J.R.: Kinematics and mobility analysis of carton folds in packing manipulation based on the mechanism equivalent. Proc. Inst. Mech. Eng. Part C: J. Mech. Eng. Sci. **216**(10), 959–970 (2002)
4. Dai, J., Caldwell, D.: Origami-based robotic paper-and-board packaging for food industry. Trends Food Sci. Technol. **21**(3), 153–157 (2010)
5. Howell, L.L.: Compliant Mechanisms. Wiley-Interscience (2001)
6. Carroll, D.W., Magleby, S.P., Howell, L.L., Todd, R.H., Lusk, C.P.: Simplified manufacturing through a metamorphic process for compliant ortho-planar mechanisms. In: ASME 2005 International Mechanical Engineering Congress and Exposition, pp. 389–399. American Society of Mechanical Engineers (2005)
7. Winder, B.G., Magleby, S.P., Howell, L.L.: Kinematic representations of pop-up paper mechanisms. J. Mech. Robot. **1**(2), 021009 (2009)
8. Song, J., Chen, Y., Lu, G.: Axial crushing of thin-walled structures with origami patterns. Thin-Walled Struct. **54**, 65–71 (2012)
9. Ma, J., You, Z.: Energy absorption of thin-walled square tubes with a prefolded origami pattern-part I: geometry and numerical simulation. J. Appl. Mech. **81**(1), 011003 (2014)
10. Zirbel, S.A., Lang, R.J., Thomson, M.W., Sigel, D.A., Walkemeyer, P.E., Trease, B.P., Magleby, S.P., Howell, L.L.: Accommodating thickness in origami-based deployable arrays. J. Mech. Des. **135**(11), 111005 (2013)
11. Chen, Y., Peng, R., You, Z.: Origami of thick panels. Science **349**(6246), 396–400 (2015)
12. Bassik, N., Stern, G.M., Gracias, D.H.: Microassembly based on hands free origami with bidirectional curvature. Appl. Phys. Lett. **95**(9), 091901 (2009)
13. McGough, K., Ahmed, S., Frecker, M., Ounaies, Z.: Finite element analysis and validation of dielectric elastomer actuators used for active origami. Smart Mater. Struct. **23**(9), 094002 (2014)
14. Lee, D.-Y., Kim, J.-S., Kim, S.-R., Koh, J.-S., Cho, K.-J.: The deformable wheel robot using magic-ball origami structure. In: Proceedings of the 2013 ASME Design Engineering Technical Conference, Portland, OR (2013)
15. Onal, C.D., Wood, R.J., Rus, D.: An origami-inspired approach to worm robots. IEEE/ASME Trans. Mechatron. **18**(2), 430–438 (2013)
16. Vander Hoff, E., Jeong, D., Lee, K.: Origamibot-i: a thread-actuated origami robot for manipulation and locomotion. In: 2014 IEEE/RSJ International Conference on Intelligent Robots and Systems (IROS 2014), pp. 1421–1426. IEEE (2014)
17. Zhang, K., Qiu, C., Dai, J.S.: Helical Kirigami-enabled centimeter-scale worm robot with shape-memory-alloy linear actuators. J. Mech. Robot. **7**(2), 021014 (2015)
18. Dai, J.S., Jones, J.R.: Mobility in metamorphic mechanisms of foldable/erectable kinds. J. Mech. Des. **121**(3), 375–382 (1999)
19. Bowen, L.A., Grames, C.L., Magleby, S.P., Howell, L.L., Lang, R.J.: A classification of action origami as systems of spherical mechanisms. J. Mech. Des. **135**(11), 111008 (2013)
20. Hull, T.: On the mathematics of flat origamis. Congr. Numer. 215–224 (1994)
21. Liu, H., Dai, J.: Carton manipulation analysis using configuration transformation. Proc. Inst. Mech. Eng. Part C: J. Mech. Eng. Sci. **216**(5), 543–555 (2002)
22. Mitani, J.: A design method for 3d origami based on rotational sweep. Comput. Aided Des. Appl. **6**(1), 69–79 (2009)
23. Dai, J.S., Wang, D., Cui, L.: Orientation and workspace analysis of the multifingered metamorphic hand-metahand. IEEE Trans. Robot. **25**(4), 942–947 (2009)
24. Wilding, S.E., Howell, L.L., Magleby, S.P.: Spherical lamina emergent mechanisms. Mech. Mach. Theory **49**, 187–197 (2012)

25. Bowen, L., Frecker, M., Simpson, T.W., von Lockette, P.: A dynamic model of magneto-active elastomer actuation of the waterbomb base. In: ASME 2014 International Design Engineering Technical Conferences and Computers and Information in Engineering Conference, pp. V05BT08A051–V05BT08A051. American Society of Mechanical Engineers (2014)
26. Yao, W., Dai, J.S.: Dexterous manipulation of origami cartons with robotic fingers based on the interactive configuration space. J. Mech. Des. **130**(2), 022303 (2008)
27. Zhang, K., Fang, Y., Fang, H., Dai, J.S.: Geometry and constraint analysis of the three-spherical kinematic chain based parallel mechanism. J. Mech. Robot. **2**(3), 031014 (2010)
28. Wei, G., Dai, J.S.: Origami-inspired integrated planar-spherical overconstrained mechanisms. J. Mech. Des. **136**(5), 051003 (2014)
29. Beex, L., Peerlings, R.: An experimental and computational study of laminated paperboard creasing and folding. Int. J. Solids Struct. **46**(24), 4192–4207 (2009)
30. Felton, S.M., Tolley, M.T., Shin, B., Onal, C.D., Demaine, E.D., Rus, D., Wood, R.J.: Self-folding with shape memory composites. Soft Matter **9**(32), 7688–7694 (2013)
31. Ahmed, S., Lauff, C., Crivaro, A., McGough, K., Sheridan, R., Frecker, M., von Lockette, P., Ounaies, Z., Simpson, T., Lien, J., et al.: Multi-field responsive origami structures: preliminary modeling and experiments. ASME Paper No. DETC2013-12405 (2013)
32. Delimont, I.L., Magleby, S.P., Howell, L.L.: Evaluating compliant hinge geometries for origami-inspired mechanisms. J. Mech. Robot. **7**(1), 011009 (2015)
33. Dai, J.S., Cannella, F.: Stiffness characteristics of carton folds for packaging. J. Mech. Des. **130**(2) (2008)
34. Qiu, C., Aminzadeh, V., Dai, J.S.: Kinematic analysis and stiffness validation of origami cartons. J. Mech. Des. **135**(11), 111004 (2013)
35. Mentrasti, L., Cannella, F., Pupilli, M., Dai, J.S.: Large bending behavior of creased paperboard. I. Experimental investigations. Int. J. Solids Struct. **50**(20), 3089–3096 (2013)
36. Mentrasti, L., Cannella, F., Pupilli, M., Dai, J.S.: Large bending behavior of creased paperboard. II. Structural analysis. Int. J. Solids Struct. **50**(20), 3097–3105 (2013)
37. Hanna, B.H., Lund, J.M., Lang, R.J., Magleby, S.P., Howell, L.L.: Waterbomb base: a symmetric single-vertex bistable origami mechanism. Smart Mater. Struct. **23**(9), 094009 (2014)
38. Hanna, B.H., Magleby, S., Lang, R.J., Howell, L.L.: Force-deflection modeling for generalized origami waterbomb-base mechanisms. J. Appl. Mech. (2015)
39. Qiu, C., Vahid, A., Dai, J.S.: Kinematic analysis and stiffness validation of origami cartons. J. Mech. Des. **135**(11), 111004 (2013)
40. Dai, J.S., Rees Jones, J.: Interrelationship between screw systems and corresponding reciprocal systems and applications. Mech. Mach. Theory **36**(5), 633–651 (2001)
41. Zhang, K., Qiu, C., Dai, J.S.: An origami-parallel structure integrated deployable continuum robot. In: Proceedings of ASME 2015 International Design Engineering Technical Conferences and Computers and Information in Engineering Conference, Boston, Massachusetts, USA, August 2–5, 2013. ASME, Boston (2015)
42. Zhang, K., Fang, Y.: Kinematics and workspace analysis of a novel spatial 3-DOF parallel manipulator. Prog. Nat. Sci. **18**(4), 432–440 (2008)
43. Murray, R.M., Li, Z., Sastry, S.S., Sastry, S.S.: A Mathematical Introduction to Robotic Manipulation. CRC Press (1994)
44. Lipkin, H., Duffy, J.: The elliptic polarity of screws. ASME J. Mech. Trans. Autom. Des. **107**, 377–387 (1985)

Chapter 9
Conclusions and Future Work

9.1 Main Achievements and Novelty

A variety of research outcomes presented in this book are considered as original work. The items listed below are given as the main achievements and novelty of this book:

1. From Chaps. 2–4, the compliance analysis and design of both traditional mechanisms and compliant mechanisms were unified in the framework of screw theory, including the description of flexible elements as well as the stiffness/compliance construction of the whole mechanisms. This unified design framework paves the way for the further systematical design of compliant mechanisms.
2. In Chap. 5, a constraint-based approach was proposed at the conceptual design level to synthesize the constraint and actuation space of compliant parallel mechanisms [1]. In this approach, the reciprocal relationship in screw theory is utilized to obtain the layout of constraints as well as actuators according to desired degrees of freedom. This led to the design of a novel compliant parallel mechanism that employs shape-memory-alloy spring based actuators and can realize orthogonal four degrees of freedom [2].
3. In Chap. 6, this compliant parallel platform was further utilized to investigate both stiffness analysis and synthesis design problems [3]. In the stiffness analysis, the layout of constraints obtained according to the reciprocal relationship with motions is utilized to construct the corresponding stiffness matrix. In the stiffness synthesis, however, the developed stiffness matrix is decomposed in return to obtain the configuration of constraint limbs based on their stiffness properties. In order to generate meaningful synthesis results, existing synthesis algorithms were compared and categorized, leading to the development of a line-vector based matrix-partition approach that can establish a one-to-one correspondence between the synthesized result and the initial configuration of constraint limbs.

© The Editor(s) (if applicable) and The Author(s), under exclusive license to Springer Nature Switzerland AG 2021
C. Qiu and J. S. Dai, *Analysis and Synthesis of Compliant Parallel Mechanisms—Screw Theory Approach*, Springer Tracts in Advanced Robotics 139,
https://doi.org/10.1007/978-3-030-48313-5_9

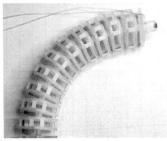
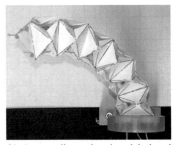

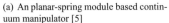

(a) An planar-spring module based continuum manipulator [5]   (b) An compliant origami-module based continuum manipulator [9]

Fig. 9.1 A series of compliant-module based continuum manipulator

4. In Chap. 7, At the dimensional design level, the optimization design of compliant mechanisms was accomplished by conducting a compliance parameterization analysis of ortho-planar springs [4]. By using a compliance-matrix based approach in the framework of screw theory, the six-dimensional compliance characteristics of ortho-planar springs were investigated analytically, leading to the further parameterization and optimization design with respect to design parameters of ortho-planar springs. It was found an ortho-planar spring not only has a large linear out-of-plane compliance but also has a large rotational bending compliance. This unique property was further utilized to develop a novel continuum manipulator that demonstrates the large bending capabilities of its planar-spring modules [5], which is shown in Fig. 9.1a.

5. In Chap. 8, the large deformation problem of compliant mechanisms was investigated using origami-inspired compliant mechanisms. Initially, experiments of single crease-type flexures were conducted to obtain their torsional stiffness performance [6]. Based on obtained stiffness properties of crease flexures, the force behaviours of compliant origami structures under large deformations were evaluated using the mechanism equivalent approach. Particularly, a novel repelling-screw based approach was developed for the first time that is able to conduct forward force analysis of compliant origami platforms [7]. Notably, a series of continuum manipulators and bio-inspired robots were further developed using the evaluated compliant origami modules [8, 9]. For example, Fig. 9.1b presents an origami-module based continuum manipulator, which demonstrates good flexibility with controllable motion in a large workspace.

9.2 Future Work

The current research establishes the feasibility of screw theory in the design of compliant mechanisms. It now remains to extend the capabilities of the present research from both theoretical developments and practical applications.

9.2.1 Theoretical Developments

9.2.1.1 Repelling Screws

A repelling-screw based force analysis approach was proposed in Chap. 8 to conduct a forward force analysis of origami compliant mechanisms. In this study, the origami compliant mechanism was equivalent to a 3-DOF parallel platform with embedded torsional stiffness in each revolute joint. A further observation indicates this repelling-screw based approach provides a geometrical interpretation of the inverse of Jacobian [10]. As such, there is a need to conduct a more comprehensive study of repelling-screws with respect to their uniqueness as well as capabilities to interpret Jacobian matrices that correspond to a wider range of mechanism configurations, either in full rank or less than full rank.

9.2.1.2 Large Deformation Problem

In Chap. 8, quasi-static force analysis was conducted to investigate the large deformation of compliant mechanisms. Particularly compliant origami mechanisms that contain crease-type flexures were selected as examples, which lies with the fact crease-type flexures can be modelled as revolute joints with fixed rotational axes and embedded rotational stiffness. Thus the complex nonlinear force/deflection behaviour of a compliant origami mechanism was decomposed into the material nonlinearities of crease flexures as well as geometrical nonlinearities of the kinematics of the equivalent mechanism, which largely simplified the large deformation analysis of the related compliant origami mechanism.

In the future, this quasi-static force analysis will be extended to compliant mechanisms with more types of flexible elements such as beam-type flexures. For example, a corotational concept [11] has been proposed to study the large deformation of slender continuous beams, which aims to separate the rigid-body motions of flexible elements from their strain producing deformations. This is exactly the same as the mechanism-equivalence principle which treats a compliant mechanism as an integration of flexible elements, but it extends from single-DOF flexible elements to multi DOF ones such as the beam-type flexures. Accordingly, this corotational formulation will be implemented into the currently developed design framework and will be utilized to design and analyze large deformation properties of compliant mechanisms with more types of flexible elements, and in both serial and parallel configurations.

9.2.2 Practical Applications

Apart from the theoretical developments, efforts can also be spent on developing compliant and soft robotic systems for applications, including but not limited to, medical tools in minimally invasive surgery, ultrasound scanning and robot-assisted rehabilitation, as well as adaptive robotic grippers for industrial manipulation.

9.2.2.1 Compliant Robot Design

The design of compliant mechanisms and robots will be continued within the design framework established in this book. Both origami and compliant-module based continuum manipulators will serve as benchmarks, based on what alternative designs will be explored by considering emerging fabrication techniques as well as novel sensing and actuation strategies, such as the silicon-body soft robot with embedded hollow chambers that are pneumatically actuated. Analytical approaches for modelling soft robot's kinematics and compliance will be extended to evaluate novel designs with various materials, configurations and actuation systems, and thus: (1) achieve optimum design of key performance criteria such as weight, workspace and compliance distribution; (2) develop control strategy for improved manipulation performance, such as providing a desired stiffness at the tip of a continuum manipulator for robot-environment interaction.

9.2.2.2 Adaptive Multi-sensing System

Multi-sensing system will be developed and integrated into compliant robots to improve both internal robotic system control and external environment interaction. From a low-level, the development of compliant structures will be continued to design integrated and distributed force/tactile sensing systems together with techniques such as optical fibres and resistive flex sensors. From a high-level, algorithms with respect to solo force/tactile sensing as well as multi-sensing fusion will be sought and implemented to improve compliant robot performance.

9.2.2.3 Actuation and Control System

The development of actuation and control system will be highly interactive with structural design of compliant robots. For the previously developed robots such as origami-inspired and compliant-module based continuum manipulators, they are extrinsically actuated by variable length tendons such as tension cables or shape-memory-alloy actuators. As such, hybrid motion and force control strategies will be developed to achieve both position and stiffness control using compliant robots' kinematics and compliance models. For potential designs of compliant robots that are actuated intrinsically, such as using pneumatic actuators, alternative control strategies will be sought. For example, antagonistic actuation principle that combines both tendon-driven and pneumatic actuation can be utilized to achieve variable stiffness control and enhance the manipulation capabilities of compliant robots.

References

1. Yu, J.J., Li, S.Z., Qiu, C.: An analytical approach for synthesizing line actuation spaces of parallel flexure mechanisms. J. Mech. Des. **135**(12), 124501 (2013)
2. Qiu, C., Zhang, K.T., Dai, J.S.: Constraint-based design and analysis of a compliant parallel mechanism using SMA-spring actuators. In: Proceedings of ASME 2014 International Design Engineering Technical Conferences and Computers and Information in Engineering Conference, New York, Buffalo, USA, August 17–20, 2014. ASME, New York (2014)
3. Qiu, C., Dai, J.S.: Constraint stiffness construction and decomposition of a SPS orthogonal parallel mechanism. In: Proceedings of ASME 2015 International Design Engineering Technical Conferences and Computers and Information in Engineering Conference, Massachusetts, Boston, USA, August 2–5, 2015. ASME (2015)
4. Qiu, C., Qi, P., Liu, H., Althoefer, K., Dai, J.S.: Six-dimensional compliance analysis and validation of ortho-planar springs. J. Mech. Des. (2016)
5. Qi, P., Qiu, C., Liu, H.B., Dai, J.S., Seneviratne, L., Althoefer, K.: A novel continuum manipulator design using serially connected double-layer planar springs. IEEE/ASME Trans. Mechatron. (2015)
6. Qiu, C., Vahid, A., Dai, J.S.: Kinematic analysis and stiffness validation of origami cartons. J. Mech. Des. **135**(11), 111004 (2013)
7. Qiu, C., Zhang, K., Dai, J.S.: Repelling-screw based force analysis of origami mechanisms. J. Mech. Robot. **15**(1122), 1 (2015)
8. Zhang, K., Qiu, C., Dai, J.S.: Helical Kirigami-enabled centimeter-scale worm robot with shape-memory-alloy linear actuators. J. Mech. Robot. **7**(2), 021014 (2015)
9. Zhang, K., Qiu, C., Dai, J.S.: An extensible continuum robot with integrated origami parallel modules. J. Mech. Robot. (2015)
10. Duffy, J.: Statics and Kinematics with Applications to Robotics. Cambridge University Press (1996)
11. Crisfield, M.A.: A consistent co-rotational formulation for non-linear, three-dimensional, beam-elements. Comput. Methods Appl. Mech. Eng. **81**(2), 131–150 (1990)

Appendix A
Constraint and Actuation Space of the Compliant Parallel Mechanism

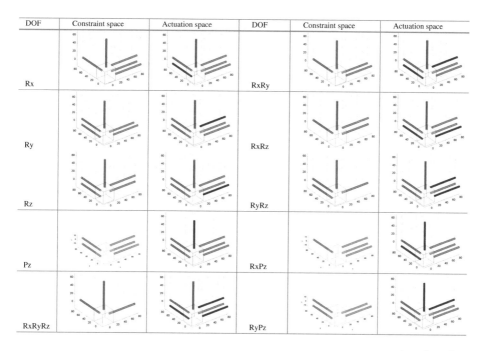

DOF	Constraint space	Actuation space	DOF	Constraint space	Actuation space
Rx			RxRy		
Ry			RxRz		
Rz			RyRz		
Pz			RxPz		
RxRyRz			RyPz		

(continued)

153

C. Qiu and J. S. Dai, *Analysis and Synthesis of Compliant Parallel Mechanisms—Screw Theory Approach*, Springer Tracts in Advanced Robotics 139, https://doi.org/10.1007/978-3-030-48313-5

Table A.1 (continued)

DOF	Constraint space	Actuation space	DOF	Constraint space	Actuation space
RxRyPz			RzPz		
RxRzPz			RxRyRzPz		
RyRzPz					

Appendix B
Selected Constraint Spaces of the Compliant Parallel Mechanism and Corresponding Decomposition Results

Rank	DOF	Constraint space	Matrix-partition	Direct-recursion
$rank(\mathbf{K}_a) = 3$ $rank(\mathbf{K}_b) = 3$	3-2-1			
$rank(\mathbf{K}_a) = 3$ $rank(\mathbf{K}_b) = 3$	2-2-2			
$rank(\mathbf{K}_a) = 3$ $rank(\mathbf{K}_b) = 2$	Rz			

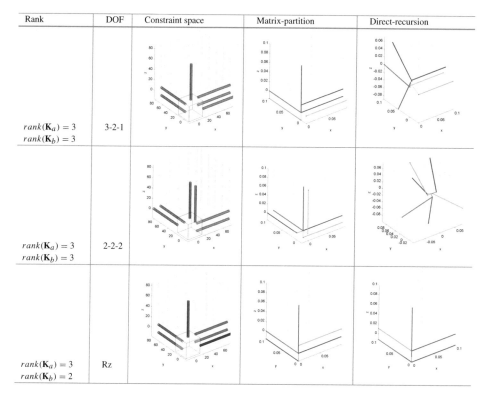

(continued)

C. Qiu and J. S. Dai, *Analysis and Synthesis of Compliant Parallel Mechanisms—Screw Theory Approach*, Springer Tracts in Advanced Robotics 139, https://doi.org/10.1007/978-3-030-48313-5

Table B.1 (continued)

Rank	DOF	Constraint space	Matrix-partition	Direct-recursion
$rank(\mathbf{K}_a) = 2$ $rank(\mathbf{K}_b) = 3$	Pz			
$rank(\mathbf{K}_a) = 2$ $rank(\mathbf{K}_b) = 2$	RyPz			

Appendix C
A Complete List of Ortho-Planar-Spring Configurations Used in FEA Simulations

A complete list of planar-spring models used in the FEA simulations is presented in Fig. C.1. The variant geometrical parameters in this list is in accordance with Table 7.2. Simulation results of these planar springs are compared with analytical values in Figs. 7.3–7.6, and the corresponding discrepancy curves are further illustrated in Fig. 7.9.

© The Editor(s) (if applicable) and The Author(s), under exclusive license to Springer Nature Switzerland AG 2021
C. Qiu and J. S. Dai, *Analysis and Synthesis of Compliant Parallel Mechanisms—Screw Theory Approach*, Springer Tracts in Advanced Robotics 139, https://doi.org/10.1007/978-3-030-48313-5

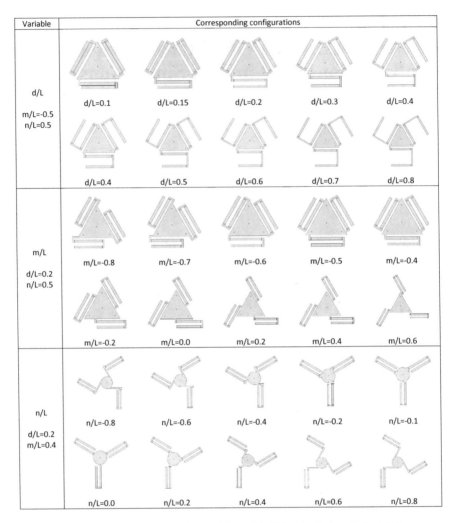

Fig. C.1 A complete list of planar-spring models used in FEM simulations in Sect. 7.4

Appendix D
Formulation of the Virtual Symmetric Plane of Waterbomb Parallel Mechanism

The equation of symmetric plane H can be determined according to the vertices A_1, A_2 and A_3. In the global coordinate frame $\{O_b, x, y, z\}$, the coordinates of A_i can be obtained using homogeneous coordinate transformation as

$$\begin{bmatrix} A_i \\ 1 \end{bmatrix} = \begin{bmatrix} \mathbf{R}_i & \mathbf{0} \\ \mathbf{0} & 1 \end{bmatrix} \begin{bmatrix} \mathbf{I} & \mathbf{P}_i \\ \mathbf{0} & 1 \end{bmatrix} \begin{bmatrix} A_i' \\ 1 \end{bmatrix}^T \tag{D.1}$$

where A_i' is the coordinate of A_i in the local coordinate frame $\{B_i, x_i, y_i, z_i\}$ and $A_i' = \begin{bmatrix} b\cos(-\theta_i) & 0 & b\sin(-\theta_i) \end{bmatrix}^T$. $\mathbf{R}_i$ is the coordinate rotation matrix from $\{O_b, x, y, z\}$ to $\{B_i, x, y, z\}$ and $\mathbf{P}_i$ is the anti-symmetric matrix representation of the coordinate translation vector $\mathbf{p}_i$, they have the forms as

$$\mathbf{R}_i = \mathbf{R}_z \left(-\frac{2\pi}{3}(i-1) \right)$$
$$\mathbf{p}_i = \begin{bmatrix} r & 0 & 0 \end{bmatrix}^T \tag{D.2}$$

The expression of A_i can be developed by substituting Eq. (D.2) into Eq. (D.1), which are used in determining the symmetric plane H with the following formula

$$\begin{vmatrix} x & y & z & 1 \\ x_{A1} & y_{A1} & y_{A1} & 1 \\ x_{A2} & y_{A2} & y_{A2} & 1 \\ x_{A3} & y_{A3} & y_{A3} & 1 \end{vmatrix} = 0 \tag{D.3}$$

Solving Eq. (D.3) we can obtain the equation of symmetric plane H as

$$A_h x + B_h y + C_h z + D_h = 0 \tag{D.4}$$

© The Editor(s) (if applicable) and The Author(s), under exclusive license to Springer Nature Switzerland AG 2021
C. Qiu and J. S. Dai, *Analysis and Synthesis of Compliant Parallel Mechanisms—Screw Theory Approach*, Springer Tracts in Advanced Robotics 139,
https://doi.org/10.1007/978-3-030-48313-5

where the values of A_h, B_h, C_h, D_h can be calculated by substituting Eqs. (D.1) and (D.2) into Eq. (D.3). The projection distance h_i between vertex E_i and the symmetric plane H can be calculated according to the equation of symmetric plane H in Eq. (D.4) and the coordinates of E_i in the global coordinate frame $\{O_b, x, y, z\}$, which has the form

$$E_i = \left[\cos\left(\tfrac{\pi}{3} - \tfrac{2\pi}{3}(i-1)\right) \quad \sin\left(\tfrac{\pi}{3} - \tfrac{2\pi}{3}(i-1)\right) \quad 0\right]^{\mathrm{T}} \tag{D.5}$$

Thus h_i can be obtained using the standard normal displacement equation as

$$h_i = \frac{A_h x_{ei} + B_h y_{ei} + C_h z_{ei} + D_h}{\sqrt{A_h^2 + B_h^2 + C_h^2}} \tag{D.6}$$

Appendix E
Motion Description of the Waterbomb Parallel Mechanism

Following the identification of virtual symmetric plane H, the motion of moving platform is evaluated using the relative motion of coordinate frame $\{O_p, x, y, z\}$ with respect to $\{O_b, x, y, z\}$. In the configuration shown in Fig. E.1, the moving platform can move in the direction that is perpendicular to the symmetric plane H. The position of $\{O_p, x, y, z\}$ can be obtained by calculating the length of vector $\overrightarrow{O_b O_p}$. $\overrightarrow{O_b O_p}$ is perpendicular to the symmetric plane H and intersects at the point M. The direction of $\overrightarrow{O_b O_p}$ can be yielded from Eq. (D.4) as

$$n_h = \begin{bmatrix} \frac{A_h}{Q_h} & \frac{B_h}{Q_h} & \frac{C_h}{Q_h} \end{bmatrix}^{\mathrm{T}} \tag{E.1}$$

where $Q_h = \pm\sqrt{A_h^2 + B_h^2 + C_h^2}$ and the sign of Q_h is determined so that n_h always points upwords. The coordinate of intersection point M can be written as

$$M = \frac{-D_h}{(A_h^2 + B_h^2 + C_h^2)} \begin{bmatrix} A_h & B_h & C_h \end{bmatrix}^{\mathrm{T}} \tag{E.2}$$

Since O_b and O_p are symmetric about plane H, the length $\|\overrightarrow{O_b O_p}\|$ is double the length of $\|\overrightarrow{O_b M}\|$, which can be calculated from Eq. (E.2) as

$$\|\overrightarrow{O_b O_p}\| = 2\|\overrightarrow{O_b M}\| = \frac{2D_h}{\sqrt{A_h^2 + B_h^2 + C_h^2}} \tag{E.3}$$

Similarly, the rotation of moving platform can be evaluated according to the direction of $\overrightarrow{O_b O_p}$. Two Eular angles ϕ and ψ are used to describe the orientation of

© The Editor(s) (if applicable) and The Author(s), under exclusive license to Springer Nature Switzerland AG 2021
C. Qiu and J. S. Dai, *Analysis and Synthesis of Compliant Parallel Mechanisms—Screw Theory Approach*, Springer Tracts in Advanced Robotics 139,
https://doi.org/10.1007/978-3-030-48313-5

Fig. E.1 A general
configuration of the
origami-enabled parallel
mechanism

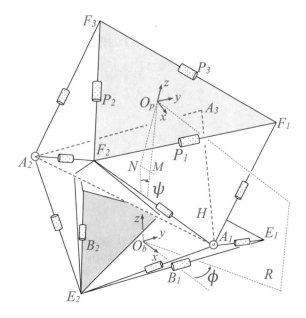

$\overrightarrow{O_b O_p}$, where ϕ is the angle between the x axis of $\{O_b, x, y, z\}$ and the projection
vector of $\overrightarrow{O_b O_p}$ on the xy plane, and ψ is the angle between the z axis and the vector
$\overrightarrow{O_b O_p}$. They can be calculated according to Eq. (E.1) as

$$\tan \phi = \frac{B_h}{A_h}$$

$$\cos \psi = \frac{C_h}{\sqrt{A_h^2 + B_h^2 + C_h^2}} \tag{E.4}$$

ϕ and ψ determine an instantaneous rotation plane R which is formed by points
O_b, O_p and N. N is the intersection point of z axis of $\{O_b, x, y, z\}$ and the symmetric
plane H, whose coordinate has the form

$$N = \begin{bmatrix} 0 & 0 & \frac{-D_h}{C_h} \end{bmatrix}^{\mathrm{T}} \tag{E.5}$$

Within the plane R, the moving platform is rotating about N and the rotation angle
is 2ψ due to the symmetry.